Week 1 Summer Practice

Real Number System

Day 1

1. Evaluate $125^{\frac{2}{3}}$. Enter your answer in the box below.

2. Which of the following expressions an equivalent radical simplified expression for $16^{\frac{5}{4}}$?

   Ⓐ $2^{\frac{5}{2}}$
   Ⓑ 32
   Ⓒ 4^5
   Ⓓ 5

3. Rewrite the radical expression $\sqrt[4]{x^3}$ as an expression with rational exponents, using the properties of exponents.

   Ⓐ $x^{\frac{4}{3}}$
   Ⓑ $x^{\frac{3}{4}}$
   Ⓒ $x^{\frac{1}{4}}$
   Ⓓ $x^{\frac{1}{12}}$

4. Rewrite the expression $y^{\frac{4}{5}}$ as an expression with rational exponents, using the properties of exponents.

   Ⓐ $\sqrt[5]{y^4}$
   Ⓑ $\sqrt[4]{y^5}$
   Ⓒ $4\sqrt[5]{y}$
   Ⓓ $\frac{\sqrt{y^4}}{5}$

5. Simplify completely $3\sqrt{18} + 2\sqrt{2}$

$9\sqrt{2} + 2\sqrt{2}$

$11\sqrt{2}$

Ⓐ $5\sqrt{20}$
Ⓑ $11\sqrt{2}$
Ⓒ $8\sqrt{2}$
Ⓓ $29\sqrt{2}$

Cite Strong and Thorough Textual Evidence

Day 1

The Open Boat

None of them knew the color of the sky. Their eyes glanced level and were fastened upon the waves that swept toward them. These waves were of the hue of slate, except for the tops, which were of foaming white, and all of the men knew the colors of the sea. The horizon narrowed and widened and dipped and rose, and at all times its edge was jagged with waves that seemed thrust up in points like rocks.

Many a man ought to have a bathtub larger than the boat which here rode upon the sea. These waves were most wrongfully and barbarously abrupt and tall, and each froth-top was a problem in small boat navigation.

The cook squatted in the bottom and looked with both eyes at the six inches of boat's side which separated him from the ocean. His sleeves were rolled over his fat forearms, and the two flaps of his unbuttoned vest dangled as he bent to bail out the boat. Often he said: "Oh! That was a narrow clip." As he remarked it, he invariably gazed eastward over the broken sea.

The oiler, steering with one of the two oars in the boat, sometimes raised himself suddenly to keep clear of water that swirled in over the stern. It was a thin little oar, and it seemed often ready to snap.

The correspondent, pulling at the other oar, watched the waves and wondered why he was there.

The injured captain, lying in the bow, was at this time buried in that profound dejection and indifference which comes, temporarily at least, to even the bravest and most enduring. When without warning the firm fails, the army loses, and the ship goes down. The mind of the master of a vessel is rooted deep in the timbers of her. Though he commanded for a day or a decade, this captain had on him the stern impression of a scene in the greys of dawn of seven turned faces. Later a stump of a top-mast with a white ball on it that slashed to and fro at the waves went low and lower and down. Thereafter there was something strange in his voice. Although steady, it was deep with mourning and of a quality beyond oration or tears.

"Keep her a little more south, Billie," said he.
"A little more south,'sir," said the oiler in the stern.

A seat in this boat was not unlike a seat upon a bucking bronco and by the same token, a bronco is not much smaller. The craft pranced and reared and plunged like an animal. As each wave came and she rose for it, she seemed like a horse making at a fence outrageously high. The manner of her scramble over these walls of water is a mystic thing, and moreover at the top of them were ordinarily these problems in white water. The foam racing down from the summit of each wave required a new leap and a leap from the air. Then, after scornfully bumping a crest, she would slide and race and splash down a long incline. It would arrive bobbing and nodding in front of the next menace.

A singular disadvantage of the sea lies in the fact that after successfully surmounting one wave, you discover that there is another behind. It is just as important and just as nervously anxious to do something effective in the way of swamping boats. In a ten-foot dinghy, one can get an idea of the resources of the sea in the line of waves that is not probable to the average experience which is never at sea in a dinghy. As each wall of water approached, it shut all else from the view of the men in the boat. It was not difficult to imagine that this particular wave was the final outburst of the ocean and the last effort of the grim water. There was a terrible grace in the move of the waves, and they came in silence, save for the snarling of the crests.

In the pale light, the faces of the men must have been grey. Their eyes must have glinted in strange ways as they gazed steadily astern. Viewed from a balcony, the whole thing would doubtlessly have been weirdly picturesque. However, the men in the boat had no time to see it. If they had leisure, there were other things to occupy their minds. The sun swung steadily up the sky, and they knew it was broad day because the color of the sea changed from slate to emerald-green, streaked with amber lights, and the foam was like tumbling snow. The process of the breaking day was unknown to them. They were aware only of this effect upon the color of the waves that rolled toward them.

Excerpt from A Letter from Joseph Conrad

To this writer of the sea, the sea was not an element. It was a stage, where displayed an exhibition of valor and of such achievement as the world had never seen before. The greatness of that achievement cannot be pronounced imaginary since its reality has affected the destinies of nations. Nevertheless, in its grandeur, it has all the remoteness of an ideal.

6. **Which of the following quotes from the story best supports the captain's feelings of the situation?**

    Ⓐ "The injured captain, lying in the bow, was at this time buried in that profound dejection and indifference…"
    Ⓑ "Although steady, it was deep with mourning and of a quality beyond oration or tears."
    Ⓒ "The mind of the master of a vessel is rooted deep in the timbers of her."
    Ⓓ "Keep her a little more south, Billie," said he.

Two Thanksgiving Day Gentlemen

Stuffy had always wondered why the Old Gentleman spoke his speech rather sadly. He did not know that it was because he was wishing every time that he had a son to succeed him. A son who would come there after he was gone. It would be a son who would stand proud and strong before some subsequent Stuffy, and say, " In memory of my father." Then it would be an Institution.

But the Old Gentleman had no relatives. He lived in rented rooms in one of the decayed old family brownstone mansions in one of the quiet streets east of the park. In the winter he raised fuchsias in a little conservatory the size of a steamer trunk. In the spring he walked in the Easter parade. In the summer he lived at a farmhouse in the New Jersey hills and sat in a wicker armchair. He would speak of a butterfly, the ornithoptera amphrisius, that he hoped to find some day. In the autumn he fed Stuffy a dinner. These were the Old Gentleman's occupations.

Stuffy Pete looked up at him for a half minute stewing and helpless in his own self-pity. The Old Gentleman's eyes were bright with the giving pleasure. His face was getting more lined each year, but his little black necktie was in as jaunty a bow as ever, and the linen was beautiful and white. His gray mustache was curled carefully at the ends. Then Stuffy made a noise that sounded like peas bubbling in a pot. Speech was intended and as the Old Gentleman had heard the sounds nine times before, he rightly construed them into Stuffy's old formula of acceptance.

"Thank you, sir. I'll go with you, and much obliged. I'm very hungry, sir."

The coma of repletion had not prevented from entering Stuffy's mind the conviction that he was the basis of an Institution. His Thanksgiving appetite was not his own. It belonged by all the sacred rights of established custom. It was not by the actual Statute of Limitations to this kind old gentleman who had preempted it. True, America is free, but in order to establish tradition, someone must be a repetend. In other words, it must be a repeating decimal. The heroes are not all heroes of steel and gold. See one here that wielded only weapons of iron, badly silvered, and tin.

The Old Gentleman led his annual protege southward to the restaurant and to the table where the feast had always occurred. They were recognized.

"Here comes the old guy," said a waiter, "he brings that same bum to a meal every Thanksgiving."

The Old Gentleman sat across the table glowing like a smoked pearl at his corner stone of future ancient Tradition. The waiters heaped the table with holiday food. Stuffy, with a sigh that was mistaken for hunger's expression, raised a knife and fork and carved for himself a crown of imperishable bay.

No more valiant hero ever fought his way through the ranks of an enemy. Turkey, chops, soups, vegetables, and pies, disappeared before him as fast as they could be served. Gorged nearly to the uttermost when he entered the restaurant, the smell of food had almost caused him to lose his honor as a gentleman. However, he rallied like a true knight. He saw the look of beneficent happiness on the Old Gentleman's face. It was a happier look than even the fuchsias and the ornithoptera amphrisius had ever brought to it. He had not the heart to see it wane.

In an hour Stuffy leaned back with a battle won. "Thank you kindly, sir," he puffed like a leaky steam pipe. "Thank you kindly for a hearty meal." Then he arose heavily with glazed eyes and started toward the kitchen. A waiter turned him about like a top and pointed him toward the door. The Old Gentleman carefully counted out $1.30 in silver change leaving three nickels for the waiter.

They parted as they did each year at the door. The Old Gentleman was going south, and Stuffy went north.

Around the first corner, Stuffy turned and stood for one minute. Then he seemed to puff out his rags as an owl puffs out his feathers and fell to the sidewalk like a sunstricken horse.

When the ambulance came the young surgeon and the driver groaned softly at his weight. Stuffy and his two dinners went to the hospital. There they stretched him on a bed and began to test him for strange diseases with the hope of getting a chance at some problem with the bare steel.

And then an hour later, another ambulance brought the Old Gentleman. They laid him on another bed and spoke of an appendicitis for he looked good for the bill.

Pretty soon one of the young doctors met one of the young nurses whose eyes he liked, and stopped to chat with her about the cases.

"That nice old gentleman over there, now," he said, "you wouldn't think that was a case of almost starvation. Proud old family, I guess. He told me he hadn't eaten a thing for three days."

7. **What quote from the passage lets the reader know that the two men have had Thanksgiving dinner together before?**

   Ⓐ "No more valiant hero ever fought his way through the ranks of an enemy."
   Ⓑ "The Old Gentleman led his annual protege southward to the restaurant and to the table where the feast had always occurred."
   Ⓒ "The heroes are not all heroes of steel and gold."
   Ⓓ "It was a happier look than even the fuchsias and the ornithoptera amphrisius had ever brought to it."

The McWilliamses and the Burglar Alarm

The conversation drifted smoothly and pleasantly along from weather to crops, from crops to literature, from literature to scandal, from scandal to religion. Then it took a random jump and landed on the subject of burglar alarms. Now for the first time Mr. McWilliams showed feeling. Whenever I perceive this sign on this man's dial, I comprehend it, lapse into silence and give him opportunity to unload his heart. He said but with ill-controlled emotion.

"I will not give one single cent for a burglar alarm, Mr. Twain—not a single cent—and I will tell you why. When we were finishing our house, we found we had a little cash left over on account of the plumber not knowing it. I was for enlarging the hearth with it because I was always unaccountably down on the hearth somehow. However, Mrs. McWilliams said, "No, let's have a burglar alarm." I agreed to this compromise. I will explain that whenever I want a thing, and Mrs. McWilliams wants another thing, and we decide upon the thing that Mrs. McWilliams wants—as we always do—she calls that a compromise. The man came up from New York and put in the alarm and charged three hundred and twenty-five dollars for it. He said we could sleep without uneasiness now. So we did for a while—say a month. Then one night we smelled smoke, and I was advised to get up and see what the matter was. I lit a candle and started toward the stairs. I met a burglar coming out of a room with a basket of our tin utensils which he had mistaken for solid silver in the dark. He was smoking a pipe.

I said, 'My friend, we do not allow smoking in this room.' He said he was a stranger and could not be expected to know the rules of the house. I said he had been in many houses just as good as this one, and it had never been objected to before. He added that as far as his experience went, such rules had never been considered to apply to burglars, anyway.

"I said, 'Smoke along, then if it is the custom. However, waiving all that, what business have you to be entering this house in this furtive and clandestine way, without ringing the burglar alarm?'

"He looked confused and ashamed and said with embarrassment, 'I beg a thousand pardons. I did not know you had a burglar alarm else I would have rung it. I beg you will not mention it where my parents may hear of it, for they are old and feeble. Such a seemingly cruel breach of the holy conventionalities of our Christian civilization might all too rudely separate the frail bridge. May I trouble you for a match?'

"I said, 'Your sentiments do you honor, but if you will allow me to say it, the metaphor is not your best hold. To return to business, how did you get in here?'

"'Through a second-story window.'

"It was even so. I redeemed the utensils at the pawnbroker's rates, less cost of advertising, bade the burglar goodnight, closed the window after him, and retired to headquarters to report. Next morning, we sent for the burglar-alarm man, and he came up and explained that the reason the alarm did not 'go off' was that no part of the house but the first floor was attached to the alarm. This was simply idiotic because one might as well have no armor on at all in battle as to have it only on his legs. The expert now put the whole second story on the alarm, charged three hundred dollars for it, and went his way. By and by, one night, I found a burglar in the third story about to start down a ladder with a lot of miscellaneous property. My first impulse was to crack his head with a billiard cue, but my second was to refrain from this attention because he was between me and the cue rack. The second impulse was plainly the soundest, so I refrained and proceeded to compromise. I redeemed the property at former rates, after deducting ten percent for use of the ladder. It was my ladder. The next day we sent down for the expert once more and had the third story attached to the alarm for three hundred dollars.

"By this time the burglar alarm had grown to difficult dimensions. It had forty-seven tags on it, marked with the names of the various rooms and chimneys, and it occupied the space of an ordinary wardrobe. The gong was the size of a wash-bowl and was placed above the head of our bed. There was a wire from the house to the coachman's quarters in the stable and a noble gong alongside his pillow.

After spending years of trying to make sure their house is secure, Mr. McWilliams would receive a bill like the one below from an expert itemizing the materials used in securing the house.

Wire	$2.15
Nipple	.75
Two hours of labor	1.50
Wax	.47
Tape	.34
Screws	.15
Recharging battery	.98
Three hours' labor	2.25
String	.02
Lard	.66
Pond's Extract	1.25
Springs at 50	2.00
Railroad fares	7.25

8. Find the evidence from the passage that supports the following idea:

Mr. McWilliams really doesn't want a burglar alarm because it's too expensive

How an Old Man Lost his Wen

Many, many years ago there lived a good old man who had a wen like a tennis-ball growing out of his right cheek. This lump was a great disfigurement to the old man, and it so annoyed him that for many years he spent all his time and money in trying to get rid of it. He tried everything he could think of. He consulted many doctors far and near and took all kinds of medicines both internally and externally. It was all of no use. The lump only grew bigger and bigger till it was nearly as big as his face, and in despair, he gave up all hopes of ever losing it. He resigned himself to the thought of having to carry the lump on his face all his life.

One day the firewood gave out in his kitchen, so as his wife wanted some at once, the old man took his ax and set out for the woods up among the hills not very far from his home. It was a fine day in the early autumn, and the old man enjoyed the fresh air and was in no hurry to get home. So, the whole afternoon passed quickly while he was chopping wood, and he had collected a goodly pile to take back to his wife. When the day began to draw to a close, he turned his face homewards.

The old man had not gone far on his way down the mountain pass when the sky clouded and rain began to fall heavily. He looked about for some shelter, but there was not even a charcoal-burner's hut near. At last, he noticed a large hole in the hollow trunk of a tree. The hole was near the ground, so he crept in easily and sat down in hopes that he had only been overtaken by a mountain shower and that the weather would soon clear.

Much to the old man's disappointment, instead of clearing the rain fell more and more heavily. Finally, a heavy thunderstorm broke over the mountain. The thunder roared so terrifically, and the heavens seemed to be so ablaze with lightning, that the old man could hardly believe himself to be alive. He thought that he must die of fright. At last, however, the sky cleared, and the whole country was aglow in the rays of the setting sun. The old man's spirits revived when he looked out at the beautiful twilight. He was about to step out from his strange hiding place in the hollow tree when the sound of what seemed like the approaching steps of several people caught his ear. He at once thought that his friends had come to look for him. He was delighted at the idea of having some jolly companions with whom to walk home. On looking out from the tree, what was his amazement to see, not his friends, but hundreds of demons coming towards the spot. The more he looked, the greater was his astonishment. Some of these demons were as large as giants. Others had great big eyes out of all proportion to the rest of their bodies. Others again had absurdly long noses, and some had such big mouths that they seemed to open from ear to ear. All had horns growing on their foreheads. The old man was so surprised at what he saw that he lost his balance and fell out of the hollow tree. Fortunately for him the demons did not see him, as the tree was in the background. So, he picked himself up and crept back into the tree.

While he was sitting there and wondering impatiently when he would be able to get home, he heard the sounds of happy music, and then some of the demons began to sing.

"What are these creatures doing?",said the old man to himself. "I will look out, as it sounds very amusing."

On peeping out, the old man saw that the demon chief himself was actually sitting with his back against the tree in which he had taken refuge, and all the other demons were sitting around, some drinking and some dancing. Food and wine were spread before them on the ground, and the demons were evidently having a great entertainment and enjoying themselves immensely.

It made the old man laugh to see their strange antics.

"How amusing this is!" laughed the old man to himself "I am now quite old, but I have never seen anything so strange in all my life."

He was so interested and excited in watching all that the demons were doing, that he forgot himself and stepped out of the tree and stood looking on.

9. **Which line from the text best supports the following idea:**

The old man didn't expect the demons to be there.

10. **Which of the quotes below best supports the following idea:**

The old man is discouraged with the lump on his face and would take more action to get it removed.

Ⓐ "He consulted many doctors far and near"
Ⓑ "there lived a good old man who had a wen like a tennis-ball growing out of his right cheek"
Ⓒ "This lump was a great disfigurement to the old man"
Ⓓ "He resigned himself to the thought of having to carry the lump on his face all his life."

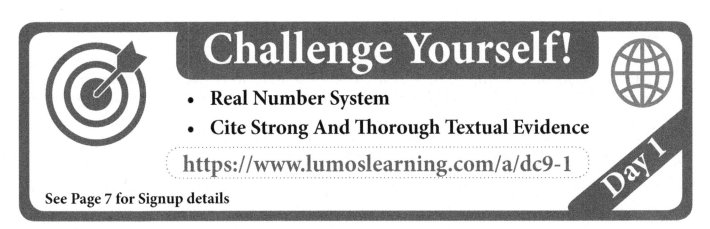

Day 2

1. As Ashley packs for her trip to Cancun, she looks up the average daily temperature for Cancun in July. She finds this is the hottest month to visit Mexico with an average daily temperature of 28°C. She knows that $F=\frac{9}{5}C+32$, where F represents degrees in Fahrenheit and C represents degrees Celsius. About what temperature should she expect in degrees Fahrenheit? Round to the nearest whole degree.

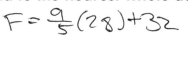

$$F=\frac{9}{5}(28)+32$$

82°

2. Alex reads about 200 words per minute. The book he needs to read is 320 pages long and has about 275 words per page. How many hours per day must he read if he has 7 days to read the entire book? Round your answer to the nearest hundredth.

3. The owner of a 1800 square foot house is replacing the flooring in the house. What is the cost to install high grade carpeting that costs $15.39 per square yard?

4. If there are 1216 students in your high school and you receive one-third of the total votes for the most favorite boy of the school then what is the reasonable number of votes you get considering the majority of students in the school voted?

   Ⓐ 405.33
   Ⓑ 398
   Ⓒ 1200
   Ⓓ 300

5. In January, Jacob is trying to predict the cost of a flight to Hawaii in July for his family vacation. Which of the following statement would give him the most accurate prediction?

   Ⓐ Asking his friend who went to Hawaii 2 years ago for Thanksgiving
   Ⓑ Finding the current cost of a ticket to Hawaii
   Ⓒ Asking his mom who went to Florida last summer
   Ⓓ Finding the cost of a ticket to Hawaii on every July 12th for the last 5 years and averaging them together

Determine A Theme Or Central Idea

The Open Boat

None of them knew the color of the sky. Their eyes glanced level and were fastened upon the waves that swept toward them. These waves were of the hue of slate, except for the tops, which were of foaming white, and all of the men knew the colors of the sea. The horizon narrowed and widened and dipped and rose, and at all times its edge was jagged with waves that seemed thrust up in points like rocks.

Many a man ought to have a bathtub larger than the boat which here rode upon the sea. These waves were most wrongfully and barbarously abrupt and tall, and each froth-top was a problem in small boat navigation.

The cook squatted in the bottom and looked with both eyes at the six inches of boat's side which separated him from the ocean. His sleeves were rolled over his fat forearms, and the two flaps of his unbuttoned vest dangled as he bent to bail out the boat. Often he said: "Oh! That was a narrow clip." As he remarked it, he invariably gazed eastward over the broken sea.

The oiler, steering with one of the two oars in the boat, sometimes raised himself suddenly to keep clear of water that swirled in over the stern. It was a thin little oar, and it seemed often ready to snap.

The correspondent, pulling at the other oar, watched the waves and wondered why he was there.

The injured captain, lying in the bow, was at this time buried in that profound dejection and indifference which comes, temporarily at least, to even the bravest and most enduring. When without warning the firm fails, the army loses, and the ship goes down. The mind of the master of a vessel is rooted deep in the timbers of her. Though he commanded for a day or a decade, this captain had on him the stern impression of a scene in the greys of dawn of seven turned faces. Later a stump of a top-mast with a white ball on it that slashed to and fro at the waves went low and lower and down. Thereafter there was something strange in his voice. Although steady, it was deep with mourning and of a quality beyond oration or tears.

"Keep her a little more south, Billie," said he.
"A little more south,'sir," said the oiler in the stern.

A seat in this boat was not unlike a seat upon a bucking bronco and by the same token, a bronco is not much smaller. The craft pranced and reared and plunged like an animal. As each wave came and she rose for it, she seemed like a horse making at a fence outrageously high. The manner of her scramble over these walls of water is a mystic thing, and moreover at the top of them were ordinarily these problems in white water. The foam racing down from the summit of each wave required a new leap and a leap from the air. Then, after scornfully bumping a crest, she would slide and race and splash down a long incline. It would arrive bobbing and nodding in front of the next menace.

A singular disadvantage of the sea lies in the fact that after successfully surmounting one wave, you discover that there is another behind. It is just as important and just as nervously anxious to do something effective in the way of swamping boats. In a ten-foot dinghy, one can get an idea of the resources of the sea in the line of waves that is not probable to the average experience which is never at sea in a dinghy. As each wall of water approached, it shut all else from the view of the men in the boat. It was not difficult to imagine that this particular wave was the final outburst of the ocean and the last effort of the grim water. There was a terrible grace in the move of the waves, and they came in silence, save for the snarling of the crests.

In the pale light, the faces of the men must have been grey. Their eyes must have glinted in strange ways as they gazed steadily astern. Viewed from a balcony, the whole thing would doubtlessly have been weirdly picturesque. However, the men in the boat had no time to see it. If they had leisure, there were other things to occupy their minds. The sun swung steadily up the sky, and they knew it was broad day because the color of the sea changed from slate to emerald-green, streaked with amber lights, and the foam was like tumbling snow. The process of the breaking day was unknown to them. They were aware only of this effect upon the color of the waves that rolled toward them.

Excerpt from A Letter from Joseph Conrad

To this writer of the sea, the sea was not an element. It was a stage, where displayed an exhibition of valor and of such achievement as the world had never seen before. The greatness of that achievement cannot be pronounced imaginary since its reality has affected the destinies of nations. Nevertheless, in its grandeur, it has all the remoteness of an ideal.

6. How does the four men's survival in the dinghy develop in the story?

Ⓐ By the captain taking charge and demanding his men comply with each task or suffer the consequences
Ⓑ By creating a chronological order of events in the plot
Ⓒ By describing how each man worked together to try to make it to shore
Ⓓ By detailing the harsh movement of the waves on the men's dinghy

Two Thanksgiving Day Gentlemen

Stuffy had always wondered why the Old Gentleman spoke his speech rather sadly. He did not know that it was because he was wishing every time that he had a son to succeed him. A son who would come there after he was gone. It would be a son who would stand proud and strong before some subsequent Stuffy, and say, " In memory of my father." Then it would be an Institution.

But the Old Gentleman had no relatives. He lived in rented rooms in one of the decayed old family brownstone mansions in one of the quiet streets east of the park. In the winter he raised fuchsias in a little conservatory the size of a steamer trunk. In the spring he walked in the Easter parade. In the summer he lived at a farmhouse in the New Jersey hills and sat in a wicker armchair. He would speak of a butterfly, the ornithoptera amphrisius, that he hoped to find some day. In the autumn he fed Stuffy a dinner. These were the Old Gentleman's occupations.

Stuffy Pete looked up at him for a half minute stewing and helpless in his own self-pity. The Old Gentleman's eyes were bright with the giving pleasure. His face was getting more lined each year, but his little black necktie was in as jaunty a bow as ever, and the linen was beautiful and white. His gray mustache was curled carefully at the ends. Then Stuffy made a noise that sounded like peas bubbling in a pot. Speech was intended and as the Old Gentleman had heard the sounds nine times before, he rightly construed them into Stuffy's old formula of acceptance.

"Thank you, sir. I'll go with you, and much obliged. I'm very hungry, sir."

The coma of repletion had not prevented from entering Stuffy's mind the conviction that he was the basis of an Institution. His Thanksgiving appetite was not his own. It belonged by all the sacred rights of established custom. It was not by the actual Statute of Limitations to this kind old gentleman who had preempted it. True, America is free, but in order to establish tradition, someone must be a repetend. In other words, it must be a repeating decimal. The heroes are not all heroes of steel and gold. See one here that wielded only weapons of iron, badly silvered, and tin.

The Old Gentleman led his annual protege southward to the restaurant and to the table where the feast had always occurred. They were recognized.

"Here comes the old guy," said a waiter, "he brings that same bum to a meal every Thanksgiving."

The Old Gentleman sat across the table glowing like a smoked pearl at his corner stone of future ancient Tradition. The waiters heaped the table with holiday food. Stuffy, with a sigh that was mistaken for hunger's expression, raised a knife and fork and carved for himself a crown of imperishable bay.

No more valiant hero ever fought his way through the ranks of an enemy. Turkey, chops, soups, vegetables, and pies, disappeared before him as fast as they could be served. Gorged nearly to the uttermost when he entered the restaurant, the smell of food had almost caused him to lose his honor as a gentleman. However, he rallied like a true knight. He saw the look of beneficent happiness on the Old Gentleman's face. It was a happier look than even the fuchsias and the ornithoptera amphrisius had ever brought to it. He had not the heart to see it wane.

In an hour Stuffy leaned back with a battle won. "Thank you kindly, sir," he puffed like a leaky steam pipe. "Thank you kindly for a hearty meal." Then he arose heavily with glazed eyes and started toward the kitchen. A waiter turned him about like a top and pointed him toward the door. The Old Gentleman carefully counted out $1.30 in silver change leaving three nickels for the waiter.

They parted as they did each year at the door. The Old Gentleman was going south, and Stuffy went north.

Around the first corner, Stuffy turned and stood for one minute. Then he seemed to puff out his rags as an owl puffs out his feathers and fell to the sidewalk like a sunstricken horse.

When the ambulance came the young surgeon and the driver groaned softly at his weight. Stuffy and his two dinners went to the hospital. There they stretched him on a bed and began to test him for strange diseases with the hope of getting a chance at some problem with the bare steel.

And then an hour later, another ambulance brought the Old Gentleman. They laid him on another bed and spoke of an appendicitis for he looked good for the bill.

Pretty soon one of the young doctors met one of the young nurses whose eyes he liked, and stopped to chat with her about the cases.

"That nice old gentleman over there, now," he said, "you wouldn't think that was a case of almost starvation. Proud old family, I guess. He told me he hadn't eaten a thing for three days."

7. How is the theme of gluttony developed in this passage?

Ⓐ By showing that eating too much can cause problems.

Ⓑ By showing that eating Thanksgiving dinner should be done together like the two characters. "The heroes are not all heroes of steel and gold."

Ⓒ The Old Gentleman and Stuffy continue a tradition despite what others think.

Ⓓ When the two gentlemen eat Thanksgiving dinner together, they realize they are not lonely.

How an Old Man Lost his Wen

Many, many years ago there lived a good old man who had a wen like a tennis-ball growing out of his right cheek. This lump was a great disfigurement to the old man, and it so annoyed him that for many years he spent all his time and money in trying to get rid of it. He tried everything he could think of. He consulted many doctors far and near and took all kinds of medicines both internally and externally. It was all of no use. The lump only grew bigger and bigger till it was nearly as big as his face, and in despair, he gave up all hopes of ever losing it. He resigned himself to the thought of having to carry the lump on his face all his life.

One day the firewood gave out in his kitchen, so as his wife wanted some at once, the old man took his ax and set out for the woods up among the hills not very far from his home. It was a fine day in the early autumn, and the old man enjoyed the fresh air and was in no hurry to get home. So, the whole afternoon passed quickly while he was chopping wood, and he had collected a goodly pile to take back to his wife. When the day began to draw to a close, he turned his face homewards.

The old man had not gone far on his way down the mountain pass when the sky clouded and rain began to fall heavily. He looked about for some shelter, but there was not even a charcoal-burner's hut near. At last, he noticed a large hole in the hollow trunk of a tree. The hole was near the ground, so he crept in easily and sat down in hopes that he had only been overtaken by a mountain shower and that the weather would soon clear.

Much to the old man's disappointment, instead of clearing the rain fell more and more heavily. Finally, a heavy thunderstorm broke over the mountain. The thunder roared so terrifically, and the heavens seemed to be so ablaze with lightning, that the old man could hardly believe himself to be alive. He thought that he must die of fright. At last, however, the sky cleared, and the whole country was aglow in the rays of the setting sun. The old man's spirits revived when he looked out at the beautiful twilight. He was about to step out from his strange hiding place in the hollow tree when the sound of what seemed like the approaching steps of several people caught his ear. He at once thought that his friends had come to look for him. He was delighted at the idea of having some jolly companions with whom to walk home. On looking out from the tree, what was his amazement to see, not his friends, but hundreds of demons coming towards the spot. The more he looked, the greater was his astonishment. Some of these demons were as large as giants. Others had great big eyes out of all proportion to the rest of their bodies. Others again had absurdly long noses, and some had such big mouths that they seemed to open from ear to ear. All had horns growing on their foreheads. The old man was so surprised at what he saw that he lost his balance and fell out of the hollow tree. Fortunately for him the demons did not see him, as the tree was in the background. So, he picked himself up and crept back into the tree.

While he was sitting there and wondering impatiently when he would be able to get home, he heard the sounds of happy music, and then some of the demons began to sing.

"What are these creatures doing?",said the old man to himself. "I will look out, as it sounds very amusing."

On peeping out, the old man saw that the demon chief himself was actually sitting with his back against the tree in which he had taken refuge, and all the other demons were sitting around, some drinking and some dancing. Food and wine were spread before them on the ground, and the demons were evidently having a great entertainment and enjoying themselves immensely.

It made the old man laugh to see their strange antics.

"How amusing this is!" laughed the old man to himself "I am now quite old, but I have never seen anything so strange in all my life."

He was so interested and excited in watching all that the demons were doing, that he forgot himself and stepped out of the tree and stood looking on.

8. **How do the demons at the end of the passage support the central idea of the man's problem?**

Havells Stories from the Odyssey

Directions: Read the rewritten excerpt from H.L . Havell's "Stories from the Odyssey" and Homer's "Odyssey." Then answer the questions.

Thence they came to the land of the Cyclopes, a rude and monstrous tribe but favored of the immortal gods by whose bounty they live. They toil not, neither do they sow, nor till the ground, but the earth of herself brings forth for them a bountiful living of wheat and barley and huge swelling clusters of the grape. Naught know they of law or civil life, but each lives in his cave on the wild mountainside, dwelling apart, careless of his neighbors with his wife and children.

It was a dark, cloudy night, and a thick mist overspread the sea when suddenly Odysseus heard the booming of breakers on a rocky shore. Before an order could be given, or any measure taken for the safety of the ships, the little fleet was caught by a strong landward current and whirled recklessly through a narrow passage between the cliffs into a land-locked harbor. Drawing their breath with relief at their wonderful escape, they beached their vessels on the level sand and lay down to wait for the day.

In the morning they found that they had been driven to the landward shore of a long island which formed a natural breakwater to a spacious bay with a narrow entrance at either end. The island was thickly covered with woods giving shelter to a multitude of wild goats, its only inhabitants. For the Cyclopes have no ships, so that the goats were left in undisturbed possession, though the place was well suited for human habitation with a deep, rich soil and plentiful springs of water.

The first care of Odysseus was to supply the crews of his vessels, which were twelve in number, with fresh meat. Armed with bows and spears, he and a picked body of men scoured the woods in search of game. They soon obtained a plentiful booty, and nine goats were assigned to each vessel with ten for that of Odysseus. So, all that day till the setting of the sun they sat and feasted on fat venison and drank of the wine which they had taken in their raid on the Thracians.

Early the next morning Odysseus manned his own galley and set forth to explore the mainland leaving the rest of the crews to await his return on the island. As they drew near the opposite shore of the bay, the mariners came in view of a gigantic cavern overshadowed by laurel-trees. Round the front of the cavern was a wide courtyard rudely fenced with huge blocks of stone and unhewn trunks of trees.

Homer's "Odyssey"

"The land of Cyclops first, a savage kind,
Nor tamed by manners, nor by laws confined:
Untaught to plant, to turn the glebe, and sow,
They all their products to free nature owe:
The soil, untill'd, a ready harvest yields,
With wheat and barley wave the golden fields;
Spontaneous wines from weighty clusters pour,
And Jove descends in each prolific shower,
By these no statues and no rights are known,
No council held, no monarch fills the throne;
But high on hills, or airy cliffs, they dwell,

Or deep in caves whose entrance leads to hell.
Each rules his race, his neighbor not his care,
Heedless of others, to his own severe.

9. How is the theme that Odysseus is starting to encounter trouble developed?

Ⓐ By showing the inhabitant's horrific appearance.
Ⓑ By explaining how the people who inhabit the land are barbaric.
Ⓒ By describing that these people grow their own food.
Ⓓ By showing that little information is known about these people because they don't associate with others.

Eleanor

I have come to see a race of people noted for vigor of fancy and ardor of passion. Men have called me mad, but the question is not yet settled whether madness is or is not the loftiest intelligence. They who dream by day are cognizant of many things which escape those who dream only by night. In their gray visions they obtain glimpses of eternity, and thrill, in awakening to find that they have been upon the verge of the great secret. They learn something of the wisdom which is of good and more of the mere knowledge which is of evil. They penetrate, however, rudderless or compassless into the vast ocean of the overwhelming light and again like the adventures of the Nubian geographer.

We will say, then, that I am mad. I grant, at least, that there are two distinct conditions of my mental existence. It is the condition of a lucid reason not to be disputed and belonging to the memory of events forming the first era of my life—and a condition of shadow and doubt, relating to the present and to the recollection of what constitutes the second great era of my being. Therefore, what I shall tell of the earlier period to what I may relate of the later time, give only such credit as may seem due or doubt it altogether.

She whom I loved in youth, and of whom I now pen calmly and distinctly these remembrances was the sole daughter of the only sister of my mother long departed. Eleonora was the name of my cousin. We had always dwelled together beneath a tropical sun in the Valley of the Many-Colored Grass. No unguided footstep ever came upon that valley, for it lay away up among a range of giant hills that hung beetling around it. It was shutting out the sunlight from its sweetest recesses. No path was flattened in its vicinity and to reach our happy home, there was need of putting back with force the foliage of many thousands of forest trees, and of crushing to death the glories of many millions of fragrant flowers. Thus, it was that we lived all alone knowing nothing of the world without the valley—I, and my cousin, and her mother.

From the dim regions beyond the mountains at the upper end of our encircled domain, there crept out a narrow and deep river, brighter than all save the eyes of Eleanor. It would be winding stealthily about in courses, and it passed away at length through a shadowy gorge among hills still dimmer than those whence it had issued. We called it the "River of Silence," for there seemed to be a hushing influence in its flow. No murmur arose from its bed and so gently it wandered along that the pearly pebbles upon which we loved to gaze far down within its bosom stirred not at all but lay in a motionless content each in its own old station shining on gloriously forever.

The margin of the river and of the many dazzling streams that glided through devious ways into its channel as well as the spaces that extended from the margins away down into the depths of the streams until they reached the bed of pebbles at the bottom. These spots, not less than the whole surface of the valley from the river to the mountains that pulled it in were carpeted all by a soft green grass, thick, short, perfectly even, and vanilla-perfumed, but so besprinkled throughout with the yellow buttercup, the white daisy, the purple violet, and the ruby-red asphodel, that its exceeding beauty spoke to our hearts in loud tones.

10. What word best describes the narrator's love for Eleanor?

Ⓐ discreet
Ⓑ tenderly
Ⓒ obsessive
Ⓓ smitten

Challenge Yourself!

- **Quantities**
- **Determine A Theme Or Central Idea**

https://www.lumoslearning.com/a/dc9-2

Day 2

See Page 7 for Signup details

Day 3

1. **What is the term with the highest degree in the expression $7x^9y-9xy^8+2x^5y^6-5xy$?**

    Ⓐ $7x^9y$
    Ⓑ $-9xy^8$
    Ⓒ $2x^5y^6$
    Ⓓ $-5xy$

2. **What is a factor of the expression $36x^2+12x+24$?**

    Ⓐ x
    Ⓑ 12
    Ⓒ 24
    Ⓓ 12x

3. **What is the coefficient of the third term the expression $5x^3y^4+7x^2y^3-6xy^2-8xy$?**

    Ⓐ 6
    Ⓑ 7
    Ⓒ -8
    Ⓓ -6

4. **What is a factor of the expression $16x^4-8x+64$? Enter your answer into the box below.**

5. **Factor the quadratic function $f(x)=9x^2+66x+21$. What are the zeros?**

    Ⓐ $\frac{1}{3}$,7
    Ⓑ $-\frac{1}{3}$,-7
    Ⓒ -3,-7
    Ⓓ -1,-7

The Open Boat

None of them knew the color of the sky. Their eyes glanced level and were fastened upon the waves that swept toward them. These waves were of the hue of slate, except for the tops, which were of foaming white, and all of the men knew the colors of the sea. The horizon narrowed and widened and dipped and rose, and at all times its edge was jagged with waves that seemed thrust up in points like rocks.

Many a man ought to have a bathtub larger than the boat which here rode upon the sea. These waves were most wrongfully and barbarously abrupt and tall, and each froth-top was a problem in small boat navigation.

The cook squatted in the bottom and looked with both eyes at the six inches of boat's side which separated him from the ocean. His sleeves were rolled over his fat forearms, and the two flaps of his unbuttoned vest dangled as he bent to bail out the boat. Often he said: "Oh! That was a narrow clip." As he remarked it, he invariably gazed eastward over the broken sea.

The oiler, steering with one of the two oars in the boat, sometimes raised himself suddenly to keep clear of water that swirled in over the stern. It was a thin little oar, and it seemed often ready to snap.

The correspondent, pulling at the other oar, watched the waves and wondered why he was there.

The injured captain, lying in the bow, was at this time buried in that profound dejection and indifference which comes, temporarily at least, to even the bravest and most enduring. When without warning the firm fails, the army loses, and the ship goes down. The mind of the master of a vessel is rooted deep in the timbers of her. Though he commanded for a day or a decade, this captain had on him the stern impression of a scene in the greys of dawn of seven turned faces. Later a stump of a top-mast with a white ball on it that slashed to and fro at the waves went low and lower and down. Thereafter there was something strange in his voice. Although steady, it was deep with mourning and of a quality beyond oration or tears.

"Keep her a little more south, Billie," said he.
"A little more south,'sir," said the oiler in the stern.

A seat in this boat was not unlike a seat upon a bucking bronco and by the same token, a bronco is not much smaller. The craft pranced and reared and plunged like an animal. As each wave came and she rose for it, she seemed like a horse making at a fence outrageously high. The manner of her scramble over these walls of water is a mystic thing, and moreover at the top of them were ordinarily these problems in white water. The foam racing down from the summit of each wave required a new leap and a leap from the air. Then, after scornfully bumping a crest, she would slide and race and splash down a long incline. It would arrive bobbing and nodding in front of the next menace.

A singular disadvantage of the sea lies in the fact that after successfully surmounting one wave, you discover that there is another behind. It is just as important and just as nervously anxious to do something effective in the way of swamping boats. In a ten-foot dinghy, one can get an idea of the resources of the sea in the line of waves that is not probable to the average experience which is never at sea in a dinghy. As each wall of water approached, it shut all else from the view of the men in the boat. It was not difficult to imagine that this particular wave was the final outburst of the ocean and the last effort of the grim water. There was a terrible grace in the move of the waves, and they came in silence, save for the snarling of the crests.

In the pale light, the faces of the men must have been grey. Their eyes must have glinted in strange ways as they gazed steadily astern. Viewed from a balcony, the whole thing would doubtlessly have been weirdly picturesque. However, the men in the boat had no time to see it. If they had leisure, there were other things to occupy their minds. The sun swung steadily up the sky, and they knew it was broad day because the color of the sea changed from slate to emerald-green, streaked with amber lights, and the foam was like tumbling snow. The process of the breaking day was unknown to them. They were aware only of this effect upon the color of the waves that rolled toward them.

Excerpt from A Letter from Joseph Conrad

To this writer of the sea, the sea was not an element. It was a stage, where displayed an exhibition of valor and of such achievement as the world had never seen before. The greatness of that achievement cannot be pronounced imaginary since its reality has affected the destinies of nations. Nevertheless, in its grandeur, it has all the remoteness of an ideal.

6. What can you tell about the interaction among the four men who are trying to survive the seas?

Ⓐ They have united to resolve the problem with the sea.
Ⓑ They have each established their own job to do as they did on the ship.
Ⓒ They seem to be upset with each other in how they are resolving their problem.
Ⓓ The captain is leading his team without receiving any input from them.

John and Mary Lamb's Version

There was a law in the city of Athens which gave to its citizens the power of compelling their daughters to marry whomsoever they pleased. For upon a daughter's refusing to marry the man her father had chosen to be her husband, the father was empowered by this law to cause her to be put to death. This is because fathers do not often desire the death of their own daughters even though they do happen to prove a little refractory. This law was seldom or never put in execution. Perhaps the young ladies of that city were not infrequently threatened by their parents with the terrors of it.

There was one instance, however, of an old man, whose name was Egeus, who actually did come before Theseus (at that time the reigning duke of Athens), to complain that his daughter Hermia, whom he had commanded to marry Demetrius, a young man of a noble Athenian family, refused to obey him because she loved another young Athenian named Lysander. Egeus demanded justice of Theseus and desired that this cruel law might be put in force against his daughter.

Hermia pleaded in excuse for her disobedience that Demetrius had formerly professed love for her dear friend Helena. Helena loved Demetrius to distraction, but this honorable reason which Hermia gave for not obeying her father's command moved not the stern Egeus.

William Shakespeare's Version

EGEUS: Happy be Theseus, our renowned Duke!
THESEUS: Thanks, good Egeus; what's the news with thee?
EGEUS: Full of vexation come I, with complaint
Against my child, my daughter Hermia.
Stand forth, Demetrius. My noble lord,
This man hath my consent to marry her.
Stand forth, Lysander. And, my gracious Duke,
This man hath bewitch'd the bosom of my child.
Thou, thou, Lysander, thou hast given her rhymes,
And interchang'd love-tokens with my child;
Thou hast by moonlight at her window sung,
With feigning voice, verses of feigning love,
And stol'n the impression of her fantasy
With bracelets of thy hair, rings, gawds, conceits,
Knacks, trifles, nosegays, sweetmeats- messengers
Of strong prevailment in unhardened youth;
With cunning hast thou filch'd my daughter's heart;
Turn'd her obedience, which is due to me,
To stubborn harshness. And, my gracious Duke,
Be it so she will not here before your Grace
Consent to marry with Demetrius,
I beg the ancient privilege of Athens:
As she is mine I may dispose of her;
Which shall be either to this gentleman

Or to her death, according to our law
Immediately provided in that case.
THESEUS: What say you, Hermia? Be advis'd, fair maid.
To you your father should be as a god;
One that compos'd your beauties; yea, and one
To whom you are but as a form in wax,
By him imprinted, and within his power
To leave the figure, or disfigure it.
Demetrius is a worthy gentleman.
HERMIA: So is Lysander.
THESEUS: In himself he is;
But, in this kind, wanting your father's voice,
The other must be held the worthier.
HERMIA: I would my father look'd but with my eyes.
THESEUS: Rather your eyes must with his judgment look.
HERMIA: I do entreat your Grace to pardon me.
I know not by what power I am made bold,

Nor how it may concern my modesty
In such a presence here to plead my thoughts;
But I beseech your Grace that I may know
The worst that may befall me in this case,
If I refuse to wed Demetrius.
THESEUS: Either to die the death, or to abjure
For ever the society of men.
Therefore, fair Hermia, question your desires,

7. **Based on the passage, what emotion did Hermia most likely develop towards her father?**

    Ⓐ Impatience
    Ⓑ Envy
    Ⓒ Fear
    Ⓓ Bitterness

Two Thanksgiving Day Gentlemen

Stuffy had always wondered why the Old Gentleman spoke his speech rather sadly. He did not know that it was because he was wishing every time that he had a son to succeed him. A son who would come there after he was gone. It would be a son who would stand proud and strong before some subsequent Stuffy, and say, " In memory of my father." Then it would be an Institution.

But the Old Gentleman had no relatives. He lived in rented rooms in one of the decayed old family brownstone mansions in one of the quiet streets east of the park. In the winter he raised fuchsias in a little conservatory the size of a steamer trunk. In the spring he walked in the Easter parade. In the summer he lived at a farmhouse in the New Jersey hills and sat in a wicker armchair. He would speak of a butterfly, the ornithoptera amphrisius, that he hoped to find some day. In the autumn he fed Stuffy a dinner. These were the Old Gentleman's occupations.

Stuffy Pete looked up at him for a half minute stewing and helpless in his own self-pity. The Old Gentleman's eyes were bright with the giving pleasure. His face was getting more lined each year, but his little black necktie was in as jaunty a bow as ever, and the linen was beautiful and white. His gray mustache was curled carefully at the ends. Then Stuffy made a noise that sounded like peas bubbling in a pot. Speech was intended and as the Old Gentleman had heard the sounds nine times before, he rightly construed them into Stuffy's old formula of acceptance.

"Thank you, sir. I'll go with you, and much obliged. I'm very hungry, sir."

The coma of repletion had not prevented from entering Stuffy's mind the conviction that he was the basis of an Institution. His Thanksgiving appetite was not his own. It belonged by all the sacred rights of established custom. It was not by the actual Statute of Limitations to this kind old gentleman who had preempted it. True, America is free, but in order to establish tradition, someone must be a repetend. In other words, it must be a repeating decimal. The heroes are not all heroes of steel and gold. See one here that wielded only weapons of iron, badly silvered, and tin.

The Old Gentleman led his annual protege southward to the restaurant and to the table where the feast had always occurred. They were recognized.

"Here comes the old guy," said a waiter, "he brings that same bum to a meal every Thanksgiving."

The Old Gentleman sat across the table glowing like a smoked pearl at his corner stone of future ancient Tradition. The waiters heaped the table with holiday food. Stuffy, with a sigh that was mistaken for hunger's expression, raised a knife and fork and carved for himself a crown of imperishable bay.

No more valiant hero ever fought his way through the ranks of an enemy. Turkey, chops, soups, vegetables, and pies, disappeared before him as fast as they could be served. Gorged nearly to the uttermost when he entered the restaurant, the smell of food had almost caused him to lose his honor as a gentleman. However, he rallied like a true knight. He saw the look of beneficent happiness on the Old Gentleman's face. It was a happier look than even the fuchsias and the ornithoptera amphrisius had ever brought to it. He had not the heart to see it wane.

In an hour Stuffy leaned back with a battle won. "Thank you kindly, sir," he puffed like a leaky steam pipe. "Thank you kindly for a hearty meal." Then he arose heavily with glazed eyes and started toward the kitchen. A waiter turned him about like a top and pointed him toward the door. The Old Gentleman carefully counted out $1.30 in silver change leaving three nickels for the waiter.

They parted as they did each year at the door. The Old Gentleman was going south, and Stuffy went north.

Around the first corner, Stuffy turned and stood for one minute. Then he seemed to puff out his rags as an owl puffs out his feathers and fell to the sidewalk like a sunstricken horse.

When the ambulance came the young surgeon and the driver groaned softly at his weight. Stuffy and his two dinners went to the hospital. There they stretched him on a bed and began to test him for strange diseases with the hope of getting a chance at some problem with the bare steel.

And then an hour later, another ambulance brought the Old Gentleman. They laid him on another bed and spoke of an appendicitis for he looked good for the bill.

Pretty soon one of the young doctors met one of the young nurses whose eyes he liked, and stopped to chat with her about the cases.

"That nice old gentleman over there, now," he said, "you wouldn't think that was a case of almost starvation. Proud old family, I guess. He told me he hadn't eaten a thing for three days."

8. How does the staff at the restaurant react to the two men eating Thanksgiving Dinner?

Ⓐ They feel sorry for them.
Ⓑ They think it is disappointing.
Ⓒ They think it is adorable.
Ⓓ They think it is special.

The McWilliamses and the Burglar Alarm

The conversation drifted smoothly and pleasantly along from weather to crops, from crops to literature, from literature to scandal, from scandal to religion. Then it took a random jump and landed on the subject of burglar alarms. Now for the first time Mr. McWilliams showed feeling. Whenever I perceive this sign on this man's dial, I comprehend it, lapse into silence and give him opportunity to unload his heart. He said but with ill-controlled emotion.

"I will not give one single cent for a burglar alarm, Mr. Twain—not a single cent—and I will tell you why. When we were finishing our house, we found we had a little cash left over on account of the plumber not knowing it. I was for enlarging the hearth with it because I was always unaccountably down on the hearth somehow. However, Mrs. McWilliams said, "No, let's have a burglar alarm." I agreed to this compromise. I will explain that whenever I want a thing, and Mrs. McWilliams wants another thing, and we decide upon the thing that Mrs. McWilliams wants—as we always do—she calls that a compromise. The man came up from New York and put in the alarm and charged three hundred and twenty-five dollars for it. He said we could sleep without uneasiness now. So we did for a while—say a month. Then one night we smelled smoke, and I was advised to get up and see what the matter was. I lit a candle and started toward the stairs. I met a burglar coming out of a room with a basket of our tin utensils which he had mistaken for solid silver in the dark. He was smoking a pipe. I said, 'My friend, we do not allow smoking in this room.' He said he was a stranger and could not be expected to know the rules of the house. I said he had been in many houses just as good as this one, and it had never been objected to before. He added that as far as his experience went, such rules had never been considered to apply to burglars, anyway.

"I said, 'Smoke along, then if it is the custom. However, waiving all that, what business have you to be entering this house in this furtive and clandestine way, without ringing the burglar alarm?'

"He looked confused and ashamed and said with embarrassment, 'I beg a thousand pardons. I did not know you had a burglar alarm else I would have rung it. I beg you will not mention it where my parents may hear of it, for they are old and feeble. Such a seemingly cruel breach of the holy conventionalities of our Christian civilization might all too rudely separate the frail bridge. May I trouble you for a match?'

"I said, 'Your sentiments do you honor, but if you will allow me to say it, the metaphor is not your best hold. To return to business, how did you get in here?'

"'Through a second-story window.'

"It was even so. I redeemed the utensils at the pawnbroker's rates, less cost of advertising, bade the burglar goodnight, closed the window after him, and retired to headquarters to report. Next morning, we sent for the burglar-alarm man, and he came up and explained that the reason the alarm did not 'go off' was that no part of the house but the first floor was attached to the alarm. This was simply idiotic because one might as well have no armor on at all in battle as to have it only on his legs. The expert now put the whole second story on the alarm, charged three hundred dollars for it, and went his way. By and by, one night, I found a burglar in the third story about to start down a ladder with a lot of miscellaneous property. My first impulse was to crack his head with a billiard cue, but my second was to refrain from this attention because he was between me and the cue rack. The second impulse was plainly the soundest, so I refrained and proceeded to compromise. I redeemed the property at former rates, after deducting ten percent for use of the ladder. It was my ladder. The next day we sent down for the expert once more and had the third story attached to the alarm for three hundred dollars.

"By this time the burglar alarm had grown to difficult dimensions. It had forty-seven tags on it, marked with the names of the various rooms and chimneys, and it occupied the space of an ordinary wardrobe. The gong was the size of a wash-bowl and was placed above the head of our bed. There was a wire from the house to the coachman's quarters in the stable and a noble gong alongside his pillow.

After spending years of trying to make sure their house is secure, Mr. McWilliams would receive a bill like the one below from an expert itemizing the materials used in securing the house.

Wire	$2.15
Nipple	.75
Two hours of labor	1.50
Wax	.47
Tape	.34
Screws	.15
Recharging battery	.98
Three hours' labor	2.25
String	.02
Lard	.66
Pond's Extract	1.25
Springs at 50	2.00
Railroad fares	7.25

9. **How does the interaction between the narrator and the burglar contribute to the theme that burglar alarms are not valuable?**

10. Match the character from the story to how they are portrayed in the passage.

	calm and relaxed	greedy	closed-minded	fearful
Mr. McWilliams	○	○	○	○
Mrs. McWilliams	○	○	○	○
Burglar	○	○	○	○
Burlgar Alarm man	○	○	○	○

Challenge Yourself!

- **Seeing Structure in Expressions**
- **Analyze the Complex Characters**

https://www.lumoslearning.com/a/dc9-3

Day 3

See Page 7 for Signup details

Day 4

1. What is the common ratio of the geometric series 2, 10, 50, 250, 1250....?

   Ⓐ -5
   Ⓑ 12
   Ⓒ 5
   Ⓓ 10

2. Which formula can be used to find the nth term of the geometric sequence 6, 2, $\frac{2}{3}$, $\frac{2}{9}$?

   Ⓐ $a_n = 6(\frac{1}{3})^{n-1}$
   Ⓑ $a_n = \frac{1}{3}(6)^{n-1}$
   Ⓒ $a_n = 6(3)^{n-1}$
   Ⓓ none of these

3. Add these polynomials. $(9x^3+2x^2-4x+1)+(-5x^3-x^2-5x+7)$

   $4x^3+x^2-9x+8$

   Ⓐ $5x^3+3x^2-9x+8$
   Ⓑ $4x^3+x^2-7x+9$
   Ⓒ $14x^3+3x^2-9x+8$
   ⬤ $4x^3+x^2-9x+8$

4. Subtract these polynomials. $(2x^3+5x^2-7x-8) - (7x^3-8x^2+5x-3)$

   $-5x^3+13x^2-12x-5$

   Ⓐ $5x^3+13x^2+12x-5$
   ⬤ $-5x^3+13x^2-12x-5$
   Ⓒ $-5x^3-13x^2-12x+5$
   Ⓓ $5x^3+13x^2-12x+5$

5. Find the remainder when $2x^3-5x+x-3$ is divided by $x-1$

   Ⓐ -3
   Ⓑ there is no remainder
   Ⓒ -11
   Ⓓ -5

The Open Boat

None of them knew the color of the sky. Their eyes glanced level and were fastened upon the waves that swept toward them. These waves were of the hue of slate, except for the tops, which were of foaming white, and all of the men knew the colors of the sea. The horizon narrowed and widened and dipped and rose, and at all times its edge was jagged with waves that seemed thrust up in points like rocks.

Many a man ought to have a bathtub larger than the boat which here rode upon the sea. These waves were most wrongfully and barbarously abrupt and tall, and each froth-top was a problem in small boat navigation.

The cook squatted in the bottom and looked with both eyes at the six inches of boat's side which separated him from the ocean. His sleeves were rolled over his fat forearms, and the two flaps of his unbuttoned vest dangled as he bent to bail out the boat. Often he said: "Oh! That was a narrow clip." As he remarked it, he invariably gazed eastward over the broken sea.

The oiler, steering with one of the two oars in the boat, sometimes raised himself suddenly to keep clear of water that swirled in over the stern. It was a thin little oar, and it seemed often ready to snap.

The correspondent, pulling at the other oar, watched the waves and wondered why he was there.

The injured captain, lying in the bow, was at this time buried in that profound dejection and indifference which comes, temporarily at least, to even the bravest and most enduring. When without warning the firm fails, the army loses, and the ship goes down. The mind of the master of a vessel is rooted deep in the timbers of her. Though he commanded for a day or a decade, this captain had on him the stern impression of a scene in the greys of dawn of seven turned faces. Later a stump of a top-mast with a white ball on it that slashed to and fro at the waves went low and lower and down. Thereafter there was something strange in his voice. Although steady, it was deep with mourning and of a quality beyond oration or tears.

"Keep her a little more south, Billie," said he.
"A little more south,'sir," said the oiler in the stern.

A seat in this boat was not unlike a seat upon a bucking bronco and by the same token, a bronco is not much smaller. The craft pranced and reared and plunged like an animal. As each wave came and she rose for it, she seemed like a horse making at a fence outrageously high. The manner of her scramble over these walls of water is a mystic thing, and moreover at the top of them were ordinarily these problems in white water. The foam racing down from the summit of each wave required a new leap and a leap from the air. Then, after scornfully bumping a crest, she would slide and race and splash down a long incline. It would arrive bobbing and nodding in front of the next menace.

A singular disadvantage of the sea lies in the fact that after successfully surmounting one wave, you discover that there is another behind. It is just as important and just as nervously anxious to do something effective in the way of swamping boats. In a ten-foot dinghy, one can get an idea of the resources of the sea in the line of waves that is not probable to the average experience which is never at sea in a dinghy. As each wall of water approached, it shut all else from the view of the men in the boat. It was not difficult to imagine that this particular wave was the final outburst of the ocean and the last effort of the grim water. There was a terrible grace in the move of the waves, and they came in silence, save for the snarling of the crests.

In the pale light, the faces of the men must have been grey. Their eyes must have glinted in strange ways as they gazed steadily astern. Viewed from a balcony, the whole thing would doubtlessly have been weirdly picturesque. However, the men in the boat had no time to see it. If they had leisure, there were other things to occupy their minds. The sun swung steadily up the sky, and they knew it was broad day because the color of the sea changed from slate to emerald-green, streaked with amber lights, and the foam was like tumbling snow. The process of the breaking day was unknown to them. They were aware only of this effect upon the color of the waves that rolled toward them.

Excerpt from A Letter from Joseph Conrad

To this writer of the sea, the sea was not an element. It was a stage, where displayed an exhibition of valor and of such achievement as the world had never seen before. The greatness of that achievement cannot be pronounced imaginary since its reality has affected the destinies of nations. Nevertheless, in its grandeur, it has all the remoteness of an ideal.

6. **What impact does imagery in the sentence below have on the passage's tone?**

   A seat in this boat was not unlike a seat upon a bucking bronco and by the same token, a bronco is not much smaller. The craft pranced and reared and plunged like an animal.

   Ⓐ When the boat began to prance, the author displayed a sensitive tone to this part of the story.
   Ⓑ The violent bucking bronco shows a serious tone due to their chaotic situation.
   Ⓒ The bronco's movement displays the sympathetic tone toward the men.
   Ⓓ With a sarcastic tone, the author uses an analogy of a bucking bronc to describe the men's feelings.

John and Mary Lamb's Version

There was a law in the city of Athens which gave to its citizens the power of compelling their daughters to marry whomsoever they pleased. For upon a daughter's refusing to marry the man her father had chosen to be her husband, the father was empowered by this law to cause her to be put to death. This is because fathers do not often desire the death of their own daughters even though they do happen to prove a little refractory. This law was seldom or never put in execution. Perhaps the young ladies of that city were not infrequently threatened by their parents with the terrors of it.

There was one instance, however, of an old man, whose name was Egeus, who actually did come before Theseus (at that time the reigning duke of Athens), to complain that his daughter Hermia, whom he had commanded to marry Demetrius, a young man of a noble Athenian family, refused to obey him because she loved another young Athenian named Lysander. Egeus demanded justice of Theseus and desired that this cruel law might be put in force against his daughter.

Hermia pleaded in excuse for her disobedience that Demetrius had formerly professed love for her dear friend Helena. Helena loved Demetrius to distraction, but this honorable reason which Hermia gave for not obeying her father's command moved not the stern Egeus.

William Shakespeare's Version

EGEUS: Happy be Theseus, our renowned Duke!
THESEUS: Thanks, good Egeus; what's the news with thee?
EGEUS: Full of vexation come I, with complaint
Against my child, my daughter Hermia.
Stand forth, Demetrius. My noble lord,
This man hath my consent to marry her.
Stand forth, Lysander. And, my gracious Duke,
This man hath bewitch'd the bosom of my child.
Thou, thou, Lysander, thou hast given her rhymes,
And interchang'd love-tokens with my child;
Thou hast by moonlight at her window sung,
With feigning voice, verses of feigning love,
And stol'n the impression of her fantasy
With bracelets of thy hair, rings, gawds, conceits,
Knacks, trifles, nosegays, sweetmeats- messengers
Of strong prevailment in unhardened youth;
With cunning hast thou filch'd my daughter's heart;
Turn'd her obedience, which is due to me,
To stubborn harshness. And, my gracious Duke,
Be it so she will not here before your Grace
Consent to marry with Demetrius,
I beg the ancient privilege of Athens:
As she is mine I may dispose of her;
Which shall be either to this gentleman

Or to her death, according to our law

Immediately provided in that case.
THESEUS: What say you, Hermia? Be advis'd, fair maid.
To you your father should be as a god;
One that compos'd your beauties; yea, and one
To whom you are but as a form in wax,
By him imprinted, and within his power
To leave the figure, or disfigure it.
Demetrius is a worthy gentleman.
HERMIA: So is Lysander.
THESEUS: In himself he is;
But, in this kind, wanting your father's voice,
The other must be held the worthier.
HERMIA: I would my father look'd but with my eyes.
THESEUS: Rather your eyes must with his judgment look.
HERMIA: I do entreat your Grace to pardon me.
I know not by what power I am made bold,
Nor how it may concern my modesty
In such a presence here to plead my thoughts;
But I beseech your Grace that I may know
The worst that may befall me in this case,
If I refuse to wed Demetrius.
THESEUS: Either to die the death, or to abjure
For ever the society of men.
Therefore, fair Hermia, question your desires,

7. **Which phrase best shows the formal tone of Shakespeare's version?**

Ⓐ Immediately provided
Ⓑ beseech your Grace
Ⓒ what's the news
Ⓓ Thanks, good

Two Thanksgiving Day Gentlemen

Stuffy had always wondered why the Old Gentleman spoke his speech rather sadly. He did not know that it was because he was wishing every time that he had a son to succeed him. A son who would come there after he was gone. It would be a son who would stand proud and strong before some subsequent Stuffy, and say, " In memory of my father." Then it would be an Institution.

But the Old Gentleman had no relatives. He lived in rented rooms in one of the decayed old family brownstone mansions in one of the quiet streets east of the park. In the winter he raised fuchsias in a little conservatory the size of a steamer trunk. In the spring he walked in the Easter parade. In the summer he lived at a farmhouse in the New Jersey hills and sat in a wicker armchair. He would speak of a butterfly, the ornithoptera amphrisius, that he hoped to find some day. In the autumn he fed Stuffy a dinner. These were the Old Gentleman's occupations.

Stuffy Pete looked up at him for a half minute stewing and helpless in his own self-pity. The Old Gentleman's eyes were bright with the giving pleasure. His face was getting more lined each year, but his little black necktie was in as jaunty a bow as ever, and the linen was beautiful and white. His gray mustache was curled carefully at the ends. Then Stuffy made a noise that sounded like peas bubbling in a pot. Speech was intended and as the Old Gentleman had heard the sounds nine times before, he rightly construed them into Stuffy's old formula of acceptance.

"Thank you, sir. I'll go with you, and much obliged. I'm very hungry, sir."

The coma of repletion had not prevented from entering Stuffy's mind the conviction that he was the basis of an Institution. His Thanksgiving appetite was not his own. It belonged by all the sacred rights of established custom. It was not by the actual Statute of Limitations to this kind old gentleman who had preempted it. True, America is free, but in order to establish tradition, someone must be a repetend. In other words, it must be a repeating decimal. The heroes are not all heroes of steel and gold. See one here that wielded only weapons of iron, badly silvered, and tin.

The Old Gentleman led his annual protege southward to the restaurant and to the table where the feast had always occurred. They were recognized.

"Here comes the old guy," said a waiter, "he brings that same bum to a meal every Thanksgiving."

The Old Gentleman sat across the table glowing like a smoked pearl at his corner stone of future ancient Tradition. The waiters heaped the table with holiday food. Stuffy, with a sigh that was mistaken for hunger's expression, raised a knife and fork and carved for himself a crown of imperishable bay.

No more valiant hero ever fought his way through the ranks of an enemy. Turkey, chops, soups, vegetables, and pies, disappeared before him as fast as they could be served. Gorged nearly to the uttermost when he entered the restaurant, the smell of food had almost caused him to lose his honor as a gentleman. However, he rallied like a true knight. He saw the look of beneficent happiness on the Old Gentleman's face. It was a happier look than even the fuchsias and the ornithoptera amphrisius had ever brought to it. He had not the heart to see it wane.

In an hour Stuffy leaned back with a battle won. "Thank you kindly, sir," he puffed like a leaky steam pipe. "Thank you kindly for a hearty meal." Then he arose heavily with glazed eyes and started toward the kitchen. A waiter turned him about like a top and pointed him toward the door. The Old Gentleman carefully counted out $1.30 in silver change leaving three nickels for the waiter.

They parted as they did each year at the door. The Old Gentleman was going south, and Stuffy went north.
Around the first corner, Stuffy turned and stood for one minute. Then he seemed to puff out his rags as an owl puffs out his feathers and fell to the sidewalk like a sunstricken horse.
When the ambulance came the young surgeon and the driver groaned softly at his weight. Stuffy and his two dinners went to the hospital. There they stretched him on a bed and began to test him for strange diseases with the hope of getting a chance at some problem with the bare steel.

And then an hour later, another ambulance brought the Old Gentleman. They laid him on another bed and spoke of an appendicitis for he looked good for the bill.

Pretty soon one of the young doctors met one of the young nurses whose eyes he liked, and stopped to chat with her about the cases.

"That nice old gentleman over there, now," he said, "you wouldn't think that was a case of almost starvation. Proud old family, I guess. He told me he hadn't eaten a thing for three days."

8. What tone is the author conveying in the following sentence from the passage?

Stuffy and his two diners went to the hospital.

Ⓐ Uneasy
Ⓑ Righteous
Ⓒ Humor
Ⓓ Apologetic

9. What is the meaning of the figurative language in the line below?

No more valiant hero ever fought his way through the ranks of an enemy.

Ⓐ The men ate as much as possible.
Ⓑ The men had fought together in a war.
Ⓒ Waiters struggled to bring enough food to the table to satisfy their customers.
Ⓓ The Old Gentlemen had a hard time paying for the entire meal.

The McWilliamses and the Burglar Alarm

The conversation drifted smoothly and pleasantly along from weather to crops, from crops to literature, from literature to scandal, from scandal to religion. Then it took a random jump and landed on the subject of burglar alarms. Now for the first time Mr. McWilliams showed feeling. Whenever I perceive this sign on this man's dial, I comprehend it, lapse into silence and give him opportunity to unload his heart. He said but with ill-controlled emotion.

"I will not give one single cent for a burglar alarm, Mr. Twain—not a single cent—and I will tell you why. When we were finishing our house, we found we had a little cash left over on account of the plumber not knowing it. I was for enlarging the hearth with it because I was always unaccountably down on the hearth somehow. However, Mrs. McWilliams said, "No, let's have a burglar alarm." I agreed to this compromise. I will explain that whenever I want a thing, and Mrs. McWilliams wants another thing, and we decide upon the thing that Mrs. McWilliams wants—as we always do—she calls that a compromise. The man came up from New York and put in the alarm and charged three hundred and twenty-five dollars for it. He said we could sleep without uneasiness now. So we did for a while—say a month. Then one night we smelled smoke, and I was advised to get up and see what the matter was. I lit a candle and started toward the stairs. I met a burglar coming out of a room with a basket of our tin utensils which he had mistaken for solid silver in the dark. He was smoking a pipe. I said, 'My friend, we do not allow smoking in this room.' He said he was a stranger and could not be expected to know the rules of the house. I said he had been in many houses just as good as this one, and it had never been objected to before. He added that as far as his experience went, such rules had never been considered to apply to burglars, anyway.

"I said, 'Smoke along, then if it is the custom. However, waiving all that, what business have you to be entering this house in this furtive and clandestine way, without ringing the burglar alarm?'

"He looked confused and ashamed and said with embarrassment, 'I beg a thousand pardons. I did not know you had a burglar alarm else I would have rung it. I beg you will not mention it where my parents may hear of it, for they are old and feeble. Such a seemingly cruel breach of the holy conventionalities of our Christian civilization might all too rudely separate the frail bridge. May I trouble you for a match?'

"I said, 'Your sentiments do you honor, but if you will allow me to say it, the metaphor is not your best hold. To return to business, how did you get in here?'

"'Through a second-story window.'
"It was even so. I redeemed the utensils at the pawnbroker's rates, less cost of advertising, bade the burglar goodnight, closed the window after him, and retired to headquarters to report. Next morning, we sent for the burglar-alarm man, and he came up and explained that the reason the alarm did not 'go off' was that no part of the house but the first floor was attached to the alarm. This was simply idiotic because one might as well have no armor on at all in battle as to have it only on his legs. The expert now put the whole second story on the alarm, charged three hundred dollars for it, and went his way. By and by, one night, I found a burglar in the third story about to start down a ladder with a lot of miscellaneous property. My first impulse was to crack his head with a billiard cue, but my second was to refrain from this attention because he was between me and the cue rack. The second impulse was plainly the soundest, so I refrained and proceeded to compromise. I redeemed the property at former rates, after deducting ten percent for use of the ladder. It was my ladder. The next day we sent down for the expert once more and had the third story attached to the alarm for three hundred dollars.

"By this time the burglar alarm had grown to difficult dimensions. It had forty-seven tags on it, marked with the names of the various rooms and chimneys, and it occupied the space of an ordinary ward-robe. The gong was the size of a wash-bowl and was placed above the head of our bed. There was a wire from the house to the coachman's quarters in the stable and a noble gong alongside his pillow.

After spending years of trying to make sure their house is secure, Mr. McWilliams would receive a bill like the one below from an expert itemizing the materials used in securing the house.

Wire	$2.15
Nipple	.75
Two hours of labor	1.50
Wax	.47
Tape	.34
Screws	.15
Recharging battery	.98
Three hours' labor	2.25
String	.02
Lard	.66

Pond's Extract	1.25
Springs at 50	2.00
Railroad fares	7.25

10. What does the following line mean?

a lot of miscellaneous property

Ⓐ the owners' most expensive jewelery
Ⓑ variety of belongings from the house
Ⓒ large pieces like furniture
Ⓓ different coins

Challenge Yourself!

- Arithmetic with Polynomials & Rational Expressions
- The Meaning of Words and Phrases

https://www.lumoslearning.com/a/dc9-4

See Page 7 for Signup details

Day 4

Day 5

1. **Find the roots of $x^2 - x - 20 = 0$.**

   Ⓐ {5, 4}
   Ⓑ {5, -4}
   Ⓒ {-5, 4}
   Ⓓ {-5, -4}

2. **What is the solution set of the equation $6x^2 - 18x - 18 = 6$**

   Ⓐ {4, 1}
   Ⓑ {-4, -1,}
   Ⓒ {-4, -1}
   Ⓓ {4, -1}

3. **How many real roots to the equation $2x^3 - 4x^2 + 3x = 1$ do/does exist?**

   Ⓐ 1
   Ⓑ 2
   Ⓒ 3
   Ⓓ 0

4. **Evaluate the function $f(x) = a^3 + 3a^2 + 2a + 9$ at $a = -3$**

   Ⓐ 3
   Ⓑ - 51
   Ⓒ 69
   Ⓓ 2

5. **Is $(x - 2)$ a factor of $(x^3 - x^2 - x - 2)$?**

   Ⓐ No
   Ⓑ Yes
   Ⓒ It is impossible to determine

Hope is the thing with feathers

"Hope" is the thing with feathers -
That perches in the soul -
And sings the tune without the words -
And never stops - at all -

And sweetest - in the Gale - is heard -
And sore must be the storm -
That could abash the little Bird
That kept so many warm -

I've heard it in the chillest land -
And on the strangest Sea -
Yet - never - in Extremity,
It asked a crumb - of me.

6. A denotative definition:

Ⓐ is a definition that evokes how the word makes you feel.
Ⓑ is the literal definition of a word.
Ⓒ cannot be found in the dictionary.
Ⓓ is another word for a synonym.

7. A connotative definition:

Ⓐ is a definition that evokes how the word makes you feel.
Ⓑ is the literal definition of a word.
Ⓒ cannot be found in the dictionary.
Ⓓ is another word for a synonym.

8. The tone of a narrator can best be understood by analyzing the

Ⓐ denotation of a word.
Ⓑ connotation of a word.
Ⓒ neither the connotation or denotation of a word.
Ⓓ None of the above.

9. **What is Dickenson comparing hope to in these lines:**

"Hope" is the thing with feathers -/ That perches in the soul -

10. **Determine whether the following statement is True or False.**

	True	False
The following quote can best be interpreted as commentary on the power of a storm that can destroy something as strong as hope: And sore must be the storm -/ That could abash the little Bird / That kept so many warm -	○	○

Challenge Yourself!

- **Arithmetic with Polynomials & Rational Expressions**
- **The Meaning of Words and Phrases**

https://www.lumoslearning.com/a/dc9-5

Day 5

See Page 7 for Signup details

How To Learn A New Language On Your Summer Break

There are many reasons to learn a new language, from being able to watch foreign movies, improve your college and job prospects, to be able to order dinner on vacation. It makes sense to want to learn a new language, but with school and assignments and exams it can be too much to take on for most of the year. That's why the upcoming summer break is the perfect time to finally start learning Japanese, or maybe improve your French before the new school year. Here are some ways to make sure you're on the right track to being multilingual this summer!

- **Decide on an online course**

You're probably sick of school textbooks by now, so the great news is that most language courses these days are designed to be fun! Most of them have designed their apps to feel like an interactive learning game, such as DuoLingo, Rosetta Stone or Babbel. These are a great first step as you can easily get started at whatever language level you're at, you can learn on-the-go, and you're not likely to get bored. And if you're competitive, you can even challenge your friends to get on the leader-board!

- **Change the language on your devices**

You would be surprised how much your brain picks up throughout the day, even when you are not actively studying. By changing the language on your phone and social media, you will be exposing yourself to new and useful phrases constantly. It's a great way to keep your mind in language-learning mode, and you won't even notice you're doing it!

- **Watch foreign TV**

Summer break is the best time to finally catch up on all those shows you've been eating to watch throughout the year, so why not throw a few foreign language titles to the list. Kids' shows are a great place to start as they are fun and generally easy to understand, but why not challenge yourself to watch a hit film in the language you're learning (it's okay if you have the subtitles on!) This is also a great way to perfect your accent.

- **Find a buddy**

Of course, the beauty of language is to be able to communicate with more people. Having a friend or community you can practice with will really help you pick up a language fast. You could call a friend who is a native speaker, or sign up to meet an international pen pal. Even if it is just another friend that is also taking on the summer language challenge, you can hold each other accountable and have some fun chatting away. Bonus: your parents probably won't be able to eavesdrop!

Week 1 - PSAT/NMSQT Prep

- Math
- Evidence Based Reading

https://www.lumoslearning.com/a/slh9-10

See Page 7 for Signup details

Weekly Fun Summer Photo Contest

Take a picture of your summer fun activity and share it on Twitter or Instagram

Use the **#SummerLearning** mention

@LumosLearning on Twitter or

@lumos.learning on Instagram

Tag friends and increase your chances of winning the contest

Participate and stand a chance to WIN $50 Amazon gift card!

Arithmetic with Polynomials & Rational Expressions (Contd.)

Day 1

1. **Expand** $(r+2)^3$

   Ⓐ r^3+8
   Ⓑ r^3+6r^2+8
   Ⓒ $r^3+6r^2+12r+8$
   Ⓓ r^3+6

2. **What is the first term in the expansion of** $(a+b)^5$?

   Ⓐ a^5
   Ⓑ a^5b^4
   Ⓒ $5a^5$
   Ⓓ a^5b^5

3. **The expression** $\dfrac{3s-6}{s-2}$ **is equivalent to**

   Ⓐ $\dfrac{-3}{2}$
   Ⓑ 3
   Ⓒ $s-2$
   Ⓓ -36

4. **Simplify:** $\dfrac{x^2+8x+12}{x^2+3x-18}$

   Ⓐ $\dfrac{x+2}{x+6}$
   Ⓑ $\dfrac{x+2}{x+3}$
   Ⓒ $\dfrac{x+2}{x-3}$
   Ⓓ $\dfrac{x+6}{x-3}$

5. **What is the least common denominator of** $\dfrac{1}{3}$, $\dfrac{2}{5x}$ **and** $\dfrac{3}{x}$?

   Ⓐ $15x^2$
   Ⓑ $15x$
   Ⓒ $8x$
   Ⓓ $9x$

Day 1

The Open Boat

None of them knew the color of the sky. Their eyes glanced level and were fastened upon the waves that swept toward them. These waves were of the hue of slate, except for the tops, which were of foaming white, and all of the men knew the colors of the sea. The horizon narrowed and widened and dipped and rose, and at all times its edge was jagged with waves that seemed thrust up in points like rocks.

Many a man ought to have a bathtub larger than the boat which here rode upon the sea. These waves were most wrongfully and barbarously abrupt and tall, and each froth-top was a problem in small boat navigation.

The cook squatted in the bottom and looked with both eyes at the six inches of boat's side which separated him from the ocean. His sleeves were rolled over his fat forearms, and the two flaps of his unbuttoned vest dangled as he bent to bail out the boat. Often he said: "Oh! That was a narrow clip." As he remarked it, he invariably gazed eastward over the broken sea.

The oiler, steering with one of the two oars in the boat, sometimes raised himself suddenly to keep clear of water that swirled in over the stern. It was a thin little oar, and it seemed often ready to snap.

The correspondent, pulling at the other oar, watched the waves and wondered why he was there.

The injured captain, lying in the bow, was at this time buried in that profound dejection and indifference which comes, temporarily at least, to even the bravest and most enduring. When without warning the firm fails, the army loses, and the ship goes down. The mind of the master of a vessel is rooted deep in the timbers of her. Though he commanded for a day or a decade, this captain had on him the stern impression of a scene in the greys of dawn of seven turned faces. Later a stump of a top-mast with a white ball on it that slashed to and fro at the waves went low and lower and down. Thereafter there was something strange in his voice. Although steady, it was deep with mourning and of a quality beyond oration or tears.

"Keep her a little more south, Billie," said he.
"A little more south,'sir," said the oiler in the stern.

A seat in this boat was not unlike a seat upon a bucking bronco and by the same token, a bronco is not much smaller. The craft pranced and reared and plunged like an animal. As each wave came and she rose for it, she seemed like a horse making at a fence outrageously high. The manner of her scramble over these walls of water is a mystic thing, and moreover at the top of them were ordinarily these problems in white water. The foam racing down from the summit of each wave required a new leap and a leap from the air. Then, after scornfully bumping a crest, she would slide and race and splash down a long incline. It would arrive bobbing and nodding in front of the next menace.

A singular disadvantage of the sea lies in the fact that after successfully surmounting one wave, you discover that there is another behind. It is just as important and just as nervously anxious to do something effective in the way of swamping boats. In a ten-foot dinghy, one can get an idea of the resources of the sea in the line of waves that is not probable to the average experience which is never at sea in a dinghy. As each wall of water approached, it shut all else from the view of the men in the boat. It was not difficult to imagine that this particular wave was the final outburst of the ocean and the last effort of the grim water. There was a terrible grace in the move of the waves, and they came in silence, save for the snarling of the crests.

In the pale light, the faces of the men must have been grey. Their eyes must have glinted in strange ways as they gazed steadily astern. Viewed from a balcony, the whole thing would doubtlessly have been weirdly picturesque. However, the men in the boat had no time to see it. If they had leisure, there were other things to occupy their minds. The sun swung steadily up the sky, and they knew it was broad day because the color of the sea changed from slate to emerald-green, streaked with amber lights, and the foam was like tumbling snow. The process of the breaking day was unknown to them. They were aware only of this effect upon the color of the waves that rolled toward them.

Excerpt from A Letter from Joseph Conrad

To this writer of the sea, the sea was not an element. It was a stage, where displayed an exhibition of valor and of such achievement as the world had never seen before. The greatness of that achievement cannot be pronounced imaginary since its reality has affected the destinies of nations. Nevertheless, in its grandeur, it has all the remoteness of an ideal.

6. **Regarding the men's encounter with surviving the ocean after a shipwreck, what did the pace of the sea's movements and the men's action create?**

    Ⓐ Mystery
    Ⓑ Tension
    Ⓒ Passion
    Ⓓ Anger

John and Mary Lamb's Version

There was a law in the city of Athens which gave to its citizens the power of compelling their daughters to marry whomsoever they pleased. For upon a daughter's refusing to marry the man her father had chosen to be her husband, the father was empowered by this law to cause her to be put to death. This is because fathers do not often desire the death of their own daughters even though they do happen to prove a little refractory. This law was seldom or never put in execution. Perhaps the young ladies of that city were not infrequently threatened by their parents with the terrors of it.

There was one instance, however, of an old man, whose name was Egeus, who actually did come before Theseus (at that time the reigning duke of Athens), to complain that his daughter Hermia, whom he had commanded to marry Demetrius, a young man of a noble Athenian family, refused to obey him because she loved another young Athenian named Lysander. Egeus demanded justice of Theseus and desired that this cruel law might be put in force against his daughter.
Hermia pleaded in excuse for her disobedience that Demetrius had formerly professed love for her dear friend Helena. Helena loved Demetrius to distraction, but this honorable reason which Hermia gave for not obeying her father's command moved not the stern Egeus.

William Shakespeare's Version

EGEUS: Happy be Theseus, our renowned Duke!
THESEUS: Thanks, good Egeus; what's the news with thee?
EGEUS: Full of vexation come I, with complaint
Against my child, my daughter Hermia.
Stand forth, Demetrius. My noble lord,
This man hath my consent to marry her.
Stand forth, Lysander. And, my gracious Duke,
This man hath bewitch'd the bosom of my child.
Thou, thou, Lysander, thou hast given her rhymes,
And interchang'd love-tokens with my child;
Thou hast by moonlight at her window sung,
With feigning voice, verses of feigning love,
And stol'n the impression of her fantasy
With bracelets of thy hair, rings, gawds, conceits,
Knacks, trifles, nosegays, sweetmeats- messengers
Of strong prevailment in unhardened youth;
With cunning hast thou filch'd my daughter's heart;
Turn'd her obedience, which is due to me,
To stubborn harshness. And, my gracious Duke,
Be it so she will not here before your Grace
Consent to marry with Demetrius,
I beg the ancient privilege of Athens:
As she is mine I may dispose of her;
Which shall be either to this gentleman

Or to her death, according to our law
Immediately provided in that case.

THESEUS: What say you, Hermia? Be advis'd, fair maid.
To you your father should be as a god;
One that compos'd your beauties; yea, and one
To whom you are but as a form in wax,
By him imprinted, and within his power
To leave the figure, or disfigure it.
Demetrius is a worthy gentleman.
HERMIA: So is Lysander.
THESEUS: In himself he is;
But, in this kind, wanting your father's voice,
The other must be held the worthier.
HERMIA: I would my father look'd but with my eyes.
THESEUS: Rather your eyes must with his judgment look.
HERMIA: I do entreat your Grace to pardon me.
I know not by what power I am made bold,
Nor how it may concern my modesty
In such a presence here to plead my thoughts;
But I beseech your Grace that I may know
The worst that may befall me in this case,
If I refuse to wed Demetrius.
THESEUS: Either to die the death, or to abjure
For ever the society of men.
Therefore, fair Hermia, question your desires,

7. What does the structure of Shakespeare's version include?

Ⓐ Dialogue and characters
Ⓑ Parallel plots
Ⓒ Flashback
Ⓓ Irony

Two Thanksgiving Day Gentlemen

Stuffy had always wondered why the Old Gentleman spoke his speech rather sadly. He did not know that it was because he was wishing every time that he had a son to succeed him. A son who would come there after he was gone. It would be a son who would stand proud and strong before some subsequent Stuffy, and say, " In memory of my father." Then it would be an Institution.

But the Old Gentleman had no relatives. He lived in rented rooms in one of the decayed old family brownstone mansions in one of the quiet streets east of the park. In the winter he raised fuchsias in a little conservatory the size of a steamer trunk. In the spring he walked in the Easter parade. In the summer he lived at a farmhouse in the New Jersey hills and sat in a wicker armchair. He would speak of a butterfly, the ornithoptera amphrisius, that he hoped to find some day. In the autumn he fed Stuffy a dinner. These were the Old Gentleman's occupations.

Stuffy Pete looked up at him for a half minute stewing and helpless in his own self-pity. The Old Gentleman's eyes were bright with the giving pleasure. His face was getting more lined each year, but his little black necktie was in as jaunty a bow as ever, and the linen was beautiful and white. His gray mustache was curled carefully at the ends. Then Stuffy made a noise that sounded like peas bubbling in a pot. Speech was intended and as the Old Gentleman had heard the sounds nine times before, he rightly construed them into Stuffy's old formula of acceptance.

"Thank you, sir. I'll go with you, and much obliged. I'm very hungry, sir."

The coma of repletion had not prevented from entering Stuffy's mind the conviction that he was the basis of an Institution. His Thanksgiving appetite was not his own. It belonged by all the sacred rights of established custom. It was not by the actual Statute of Limitations to this kind old gentleman who had preempted it. True, America is free, but in order to establish tradition, someone must be a repetend. In other words, it must be a repeating decimal. The heroes are not all heroes of steel and gold. See one here that wielded only weapons of iron, badly silvered, and tin.
The Old Gentleman led his annual protege southward to the restaurant and to the table where the feast had always occurred. They were recognized.

"Here comes the old guy," said a waiter, "he brings that same bum to a meal every Thanksgiving."

The Old Gentleman sat across the table glowing like a smoked pearl at his corner stone of future ancient Tradition. The waiters heaped the table with holiday food. Stuffy, with a sigh that was mistaken for hunger's expression, raised a knife and fork and carved for himself a crown of imperishable bay.

No more valiant hero ever fought his way through the ranks of an enemy. Turkey, chops, soups, vegetables, and pies, disappeared before him as fast as they could be served. Gorged nearly to the uttermost when he entered the restaurant, the smell of food had almost caused him to lose his honor as a gentleman. However, he rallied like a true knight. He saw the look of beneficent happiness on the Old Gentleman's face. It was a happier look than even the fuchsias and the ornithoptera amphrisius had ever brought to it. He had not the heart to see it wane.

In an hour Stuffy leaned back with a battle won. "Thank you kindly, sir," he puffed like a leaky steam pipe. "Thank you kindly for a hearty meal." Then he arose heavily with glazed eyes and started toward the kitchen. A waiter turned him about like a top and pointed him toward the door. The Old Gentleman carefully counted out $1.30 in silver change leaving three nickels for the waiter.

They parted as they did each year at the door. The Old Gentleman was going south, and Stuffy went north.

Around the first corner, Stuffy turned and stood for one minute. Then he seemed to puff out his rags as an owl puffs out his feathers and fell to the sidewalk like a sunstricken horse.

When the ambulance came the young surgeon and the driver groaned softly at his weight. Stuffy and his two dinners went to the hospital. There they stretched him on a bed and began to test him for strange diseases with the hope of getting a chance at some problem with the bare steel.

And then an hour later, another ambulance brought the Old Gentleman. They laid him on another bed and spoke of an appendicitis for he looked good for the bill.

Pretty soon one of the young doctors met one of the young nurses whose eyes he liked, and stopped to chat with her about the cases.

"That nice old gentleman over there, now," he said, "you wouldn't think that was a case of almost starvation. Proud old family, I guess. He told me he hadn't eaten a thing for three days."

8. What element is found in the passage above?

    Ⓐ A Parallel plot
    Ⓑ A Flashback
    Ⓒ A Soliloquy
    Ⓓ An Allusion

9. When the two men go to the hospital, what type of literary device takes place during that point of the story?

    Ⓐ Dramatic Irony
    Ⓑ Sarcasm
    Ⓒ Verbal Irony
    Ⓓ Situational Irony

The McWilliamses and the Burglar Alarm

The conversation drifted smoothly and pleasantly along from weather to crops, from crops to literature, from literature to scandal, from scandal to religion. Then it took a random jump and landed on the subject of burglar alarms. Now for the first time Mr. McWilliams showed feeling. Whenever I perceive this sign on this man's dial, I comprehend it, lapse into silence and give him opportunity to unload his heart. He said but with ill-controlled emotion.

"I will not give one single cent for a burglar alarm, Mr. Twain—not a single cent—and I will tell you why. When we were finishing our house, we found we had a little cash left over on account of the plumber not knowing it. I was for enlarging the hearth with it because I was always unaccountably down on the hearth somehow. However, Mrs. McWilliams said, "No, let's have a burglar alarm." I agreed to this compromise. I will explain that whenever I want a thing, and Mrs. McWilliams wants another thing, and we decide upon the thing that Mrs. McWilliams wants—as we always do—she calls that a compromise. The man came up from New York and put in the alarm and charged three hundred and twenty-five dollars for it. He said we could sleep without uneasiness now. So we did for a while—say a month. Then one night we smelled smoke, and I was advised to get up and see what the matter was. I lit a candle and started toward the stairs. I met a burglar coming out of a room with a basket of our tin utensils which he had mistaken for solid silver in the dark. He was smoking a pipe. I said, 'My friend, we do not allow smoking in this room.' He said he was a stranger and could not be expected to know the rules of the house. I said he had been in many houses just as good as this one, and it had never been objected to before. He added that as far as his experience went, such rules had never been considered to apply to burglars, anyway.

"I said, 'Smoke along, then if it is the custom. However, waiving all that, what business have you to be entering this house in this furtive and clandestine way, without ringing the burglar alarm?'

"He looked confused and ashamed and said with embarrassment, 'I beg a thousand pardons. I did not know you had a burglar alarm else I would have rung it. I beg you will not mention it where my parents may hear of it, for they are old and feeble. Such a seemingly cruel breach of the holy conventionalities of our Christian civilization might all too rudely separate the frail bridge. May I trouble you for a match?'

"I said, 'Your sentiments do you honor, but if you will allow me to say it, the metaphor is not your best hold. To return to business, how did you get in here?'

"'Through a second-story window.'

"It was even so. I redeemed the utensils at the pawnbroker's rates, less cost of advertising, bade the burglar goodnight, closed the window after him, and retired to headquarters to report. Next morning, we sent for the burglar-alarm man, and he came up and explained that the reason the alarm did not 'go off' was that no part of the house but the first floor was attached to the alarm. This was simply idiotic because one might as well have no armor on at all in battle as to have it only on his legs. The expert now put the whole second story on the alarm, charged three hundred dollars for it, and went his way. By and by, one night, I found a burglar in the third story about to start down a ladder with a lot of miscellaneous property. My first impulse was to crack his head with a billiard cue, but my second was to refrain from this attention because he was between me and the cue rack. The second impulse was plainly the soundest, so I refrained and proceeded to compromise. I redeemed the property at former rates, after deducting ten percent for use of the ladder. It was my ladder. The next day we sent down for the expert once more and had the third story attached to the alarm for three hundred dollars.

"By this time the burglar alarm had grown to difficult dimensions. It had forty-seven tags on it, marked with the names of the various rooms and chimneys, and it occupied the space of an ordinary wardrobe. The gong was the size of a wash-bowl and was placed above the head of our bed. There was a wire from the house to the coachman's quarters in the stable and a noble gong alongside his pillow.

After spending years of trying to make sure their house is secure, Mr. McWilliams would receive a bill like the one below from an expert itemizing the materials used in securing the house.

Wire	$2.15
Nipple	.75
Two hours of labor	1.50
Wax	.47
Tape	.34
Screws	.15
Recharging battery	.98
Three hours' labor	2.25
String	.02
Lard	.66

Pond's Extract	1.25
Springs at 50	2.00
Railroad fares	7.25

10. What information from the passage foreshadows that the McWilliams will experience a second burglar?

Interpreting Functions

1. The expression $\dfrac{a}{b} - \dfrac{1}{2}$ is equivalent to

    Ⓐ $\dfrac{a-1}{b-2}$

    Ⓑ $\dfrac{a-1}{2b}$

    Ⓒ $\dfrac{2a-b}{2b}$

    Ⓓ $\dfrac{2a-b}{b-2}$

2. Simplify: $\dfrac{x^2-1}{x+2}$ and $\dfrac{x+2}{2x-2}$

    Ⓐ $\dfrac{x+1}{2(x+2)}$

    Ⓑ $\dfrac{x+2}{2(x+2)}$

    Ⓒ $\dfrac{x+1}{2}$

    Ⓓ $\dfrac{x}{2}$

3. Is the following relation a function? Answer Yes or No.

x	y
-1	5
0	9
3	6
-1	2
5	1

4. Given the ordered pairs $(-5, 2)(-4, -1)(-3, -2)(-2, -1)(-1, 2)$, identify the domain.

5. **Suppose the function g(x) doubles the square of the input and then subtracts 13. What is g(3)?**

Ⓐ 22
Ⓑ -4
Ⓒ 31
Ⓓ 5

Analyze a Particular Point of View

Day 2

The Open Boat

None of them knew the color of the sky. Their eyes glanced level and were fastened upon the waves that swept toward them. These waves were of the hue of slate, except for the tops, which were of foaming white, and all of the men knew the colors of the sea. The horizon narrowed and widened and dipped and rose, and at all times its edge was jagged with waves that seemed thrust up in points like rocks.

Many a man ought to have a bathtub larger than the boat which here rode upon the sea. These waves were most wrongfully and barbarously abrupt and tall, and each froth-top was a problem in small boat navigation.

The cook squatted in the bottom and looked with both eyes at the six inches of boat's side which separated him from the ocean. His sleeves were rolled over his fat forearms, and the two flaps of his unbuttoned vest dangled as he bent to bail out the boat. Often he said: "Oh! That was a narrow clip." As he remarked it, he invariably gazed eastward over the broken sea.

The oiler, steering with one of the two oars in the boat, sometimes raised himself suddenly to keep clear of water that swirled in over the stern. It was a thin little oar, and it seemed often ready to snap.

The correspondent, pulling at the other oar, watched the waves and wondered why he was there.

The injured captain, lying in the bow, was at this time buried in that profound dejection and indifference which comes, temporarily at least, to even the bravest and most enduring. When without warning the firm fails, the army loses, and the ship goes down. The mind of the master of a vessel is rooted deep in the timbers of her. Though he commanded for a day or a decade, this captain had on him the stern impression of a scene in the greys of dawn of seven turned faces. Later a stump of a top-mast with a white ball on it that slashed to and fro at the waves went low and lower and down. Thereafter there was something strange in his voice. Although steady, it was deep with mourning and of a quality beyond oration or tears.

"Keep her a little more south, Billie," said he.
"A little more south,'sir," said the oiler in the stern.

A seat in this boat was not unlike a seat upon a bucking bronco and by the same token, a bronco is not much smaller. The craft pranced and reared and plunged like an animal. As each wave came and she rose for it, she seemed like a horse making at a fence outrageously high. The manner of her scramble over these walls of water is a mystic thing, and moreover at the top of them were ordinarily these problems in white water. The foam racing down from the summit of each wave required a new leap and a leap from the air. Then, after scornfully bumping a crest, she would slide and race and splash down a long incline. It would arrive bobbing and nodding in front of the next menace.

A singular disadvantage of the sea lies in the fact that after successfully surmounting one wave, you discover that there is another behind. It is just as important and just as nervously anxious to do something effective in the way of swamping boats. In a ten-foot dinghy, one can get an idea of the resources of the sea in the line of waves that is not probable to the average experience which is never at sea in a dinghy. As each wall of water approached, it shut all else from the view of the men in the boat. It was not difficult to imagine that this particular wave was the final outburst of the ocean and the last effort of the grim water. There was a terrible grace in the move of the waves, and they came in silence, save for the snarling of the crests.

In the pale light, the faces of the men must have been grey. Their eyes must have glinted in strange ways as they gazed steadily astern. Viewed from a balcony, the whole thing would doubtlessly have been weirdly picturesque. However, the men in the boat had no time to see it. If they had leisure, there were other things to occupy their minds. The sun swung steadily up the sky, and they knew it was broad day because the color of the sea changed from slate to emerald-green, streaked with amber lights, and the foam was like tumbling snow. The process of the breaking day was unknown to them. They were aware only of this effect upon the color of the waves that rolled toward them.

Excerpt from A Letter from Joseph Conrad

To this writer of the sea, the sea was not an element. It was a stage, where displayed an exhibition of valor and of such achievement as the world had never seen before. The greatness of that achievement cannot be pronounced imaginary since its reality has affected the destinies of nations. Nevertheless, in its grandeur, it has all the remoteness of an ideal.

6. **The author Stephen Crane wrote this story after his encounter as a news reporter on a sinking ship. He and three others tried to survive the open sea after their ship sunk. Through which character in the story can you see Stephen Crane's point of view?**

Ⓐ The Captain, because as a writer he is taking command of the story
Ⓑ The Correspondent, because as a news reporter he represented every man
Ⓒ The Oiler, because he was the most experienced man on the ship
Ⓓ The Cook, because he was most likely the most intimidated by the sea

John and Mary Lamb's Version

There was a law in the city of Athens which gave to its citizens the power of compelling their daughters to marry whomsoever they pleased. For upon a daughter's refusing to marry the man her father had chosen to be her husband, the father was empowered by this law to cause her to be put to death. This is because fathers do not often desire the death of their own daughters even though they do happen to prove a little refractory. This law was seldom or never put in execution. Perhaps the young ladies of that city were not infrequently threatened by their parents with the terrors of it.

There was one instance, however, of an old man, whose name was Egeus, who actually did come before Theseus (at that time the reigning duke of Athens), to complain that his daughter Hermia, whom he had commanded to marry Demetrius, a young man of a noble Athenian family, refused to obey him because she loved another young Athenian named Lysander. Egeus demanded justice of Theseus and desired that this cruel law might be put in force against his daughter.

Hermia pleaded in excuse for her disobedience that Demetrius had formerly professed love for her dear friend Helena. Helena loved Demetrius to distraction, but this honorable reason which Hermia gave for not obeying her father's command moved not the stern Egeus.

William Shakespeare's Version

EGEUS: Happy be Theseus, our renowned Duke!
THESEUS: Thanks, good Egeus; what's the news with thee?
EGEUS: Full of vexation come I, with complaint
Against my child, my daughter Hermia.
Stand forth, Demetrius. My noble lord,
This man hath my consent to marry her.
Stand forth, Lysander. And, my gracious Duke,
This man hath bewitch'd the bosom of my child.
Thou, thou, Lysander, thou hast given her rhymes,
And interchang'd love-tokens with my child;
Thou hast by moonlight at her window sung,
With feigning voice, verses of feigning love,
And stol'n the impression of her fantasy
With bracelets of thy hair, rings, gawds, conceits,
Knacks, trifles, nosegays, sweetmeats- messengers
Of strong prevailment in unhardened youth;
With cunning hast thou filch'd my daughter's heart;
Turn'd her obedience, which is due to me,
To stubborn harshness. And, my gracious Duke,
Be it so she will not here before your Grace
Consent to marry with Demetrius,
I beg the ancient privilege of Athens:
As she is mine I may dispose of her;
Which shall be either to this gentleman

Or to her death, according to our law

Immediately provided in that case.
THESEUS: What say you, Hermia? Be advis'd, fair maid.
To you your father should be as a god;
One that compos'd your beauties; yea, and one
To whom you are but as a form in wax,
By him imprinted, and within his power
To leave the figure, or disfigure it.
Demetrius is a worthy gentleman.
HERMIA: So is Lysander.
THESEUS: In himself he is;
But, in this kind, wanting your father's voice,
The other must be held the worthier.
HERMIA: I would my father look'd but with my eyes.
THESEUS: Rather your eyes must with his judgment look.
HERMIA: I do entreat your Grace to pardon me.
I know not by what power I am made bold,
Nor how it may concern my modesty
In such a presence here to plead my thoughts;
But I beseech your Grace that I may know
The worst that may befall me in this case,
If I refuse to wed Demetrius.
THESEUS: Either to die the death, or to abjure
For ever the society of men.
Therefore, fair Hermia, question your desires,

7. **Which cultural problem is shown in both excerpts of "A Midsummer Night's Dream"?**

Ⓐ Child abuse
Ⓑ Immigration rights
Ⓒ Court procedures
Ⓓ Women's rights

The McWilliamses and the Burglar Alarm

The conversation drifted smoothly and pleasantly along from weather to crops, from crops to literature, from literature to scandal, from scandal to religion. Then it took a random jump and landed on the subject of burglar alarms. Now for the first time Mr. McWilliams showed feeling. Whenever I perceive this sign on this man's dial, I comprehend it, lapse into silence and give him opportunity to unload his heart. He said but with ill-controlled emotion.

"I will not give one single cent for a burglar alarm, Mr. Twain—not a single cent—and I will tell you why. When we were finishing our house, we found we had a little cash left over on account of the plumber not knowing it. I was for enlarging the hearth with it because I was always unaccountably down on the hearth somehow. However, Mrs. McWilliams said, "No, let's have a burglar alarm." I agreed to this compromise. I will explain that whenever I want a thing, and Mrs. McWilliams wants another thing, and we decide upon the thing that Mrs. McWilliams wants—as we always do—she calls that a compromise. The man came up from New York and put in the alarm and charged three hundred and twenty-five dollars for it. He said we could sleep without uneasiness now. So we did for a while—say a month. Then one night we smelled smoke, and I was advised to get up and see what the matter was. I lit a candle and started toward the stairs. I met a burglar coming out of a room with a basket of our tin utensils which he had mistaken for solid silver in the dark. He was smoking a pipe. I said, 'My friend, we do not allow smoking in this room.' He said he was a stranger and could not be expected to know the rules of the house. I said he had been in many houses just as good as this one, and it had never been objected to before. He added that as far as his experience went, such rules had never been considered to apply to burglars, anyway.

"I said, 'Smoke along, then if it is the custom. However, waiving all that, what business have you to be entering this house in this furtive and clandestine way, without ringing the burglar alarm?'

"He looked confused and ashamed and said with embarrassment, 'I beg a thousand pardons. I did not know you had a burglar alarm else I would have rung it. I beg you will not mention it where my parents may hear of it, for they are old and feeble. Such a seemingly cruel breach of the holy conventionalities of our Christian civilization might all too rudely separate the frail bridge. May I trouble you for a match?'

"I said, 'Your sentiments do you honor, but if you will allow me to say it, the metaphor is not your best hold. To return to business, how did you get in here?'

"'Through a second-story window.'

"It was even so. I redeemed the utensils at the pawnbroker's rates, less cost of advertising, bade the burglar goodnight, closed the window after him, and retired to headquarters to report. Next morning, we sent for the burglar-alarm man, and he came up and explained that the reason the alarm did not 'go off' was that no part of the house but the first floor was attached to the alarm. This was simply idiotic because one might as well have no armor on at all in battle as to have it only on his legs. The expert now put the whole second story on the alarm, charged three hundred dollars for it, and went his way. By and by, one night, I found a burglar in the third story about to start down a ladder with a lot of miscellaneous property. My first impulse was to crack his head with a billiard cue, but my second was to refrain from this attention because he was between me and the cue rack. The second impulse was plainly the soundest, so I refrained and proceeded to compromise. I redeemed the property at former rates, after deducting ten percent for use of the ladder. It was my ladder. The next day we sent down for the expert once more and had the third story attached to the alarm for three hundred dollars.

"By this time the burglar alarm had grown to difficult dimensions. It had forty-seven tags on it, marked with the names of the various rooms and chimneys, and it occupied the space of an ordinary wardrobe. The gong was the size of a wash-bowl and was placed above the head of our bed. There was a wire from the house to the coachman's quarters in the stable and a noble gong alongside his pillow.

After spending years of trying to make sure their house is secure, Mr. McWilliams would receive a bill like the one below from an expert itemizing the materials used in securing the house.

Wire	$2.15
Nipple	.75
Two hours of labor	1.50
Wax	.47
Tape	.34
Screws	.15
Recharging battery	.98
Three hours' labor	2.25
String	.02
Lard	.66
Pond's Extract	1.25
Springs at 50	2.00
Railroad fares	7.25

8. The narrator's point of view concerning the effectiveness of the burglar alarm can be described as

9. Which phrase below reveals that the passage was during a different time or place?

Ⓐ not a single cent
Ⓑ billiard cue
Ⓒ burglar alarm
Ⓓ coachman's quarters

How an Old Man Lost his Wen

Many, many years ago there lived a good old man who had a wen like a tennis-ball growing out of his right cheek. This lump was a great disfigurement to the old man, and it so annoyed him that for many years he spent all his time and money in trying to get rid of it. He tried everything he could think of. He consulted many doctors far and near and took all kinds of medicines both internally and externally. It was all of no use. The lump only grew bigger and bigger till it was nearly as big as his face, and in despair, he gave up all hopes of ever losing it. He resigned himself to the thought of having to carry the lump on his face all his life.

One day the firewood gave out in his kitchen, so as his wife wanted some at once, the old man took his ax and set out for the woods up among the hills not very far from his home. It was a fine day in the early autumn, and the old man enjoyed the fresh air and was in no hurry to get home. So, the whole afternoon passed quickly while he was chopping wood, and he had collected a goodly pile to take back to his wife. When the day began to draw to a close, he turned his face homewards.

The old man had not gone far on his way down the mountain pass when the sky clouded and rain began to fall heavily. He looked about for some shelter, but there was not even a charcoal-burner's hut near. At last, he noticed a large hole in the hollow trunk of a tree. The hole was near the ground, so he crept in easily and sat down in hopes that he had only been overtaken by a mountain shower and that the weather would soon clear.

Much to the old man's disappointment, instead of clearing the rain fell more and more heavily. Finally, a heavy thunderstorm broke over the mountain. The thunder roared so terrifically, and the heavens seemed to be so ablaze with lightning, that the old man could hardly believe himself to be alive. He thought that he must die of fright. At last, however, the sky cleared, and the whole country was aglow in the rays of the setting sun. The old man's spirits revived when he looked out at the beautiful twilight. He was about to step out from his strange hiding place in the hollow tree when the sound of what seemed like the approaching steps of several people caught his ear. He at once thought that his friends had come to look for him. He was delighted at the idea of having some jolly companions with whom to walk home. On looking out from the tree, what was his amazement to see, not his friends, but hundreds of demons coming towards the spot. The more he looked, the greater was his astonishment. Some of these demons were as large as giants. Others had great big eyes out of all proportion to the rest of their bodies. Others again had absurdly long noses, and some had such big mouths that they seemed to open from ear to ear. All had horns growing on their foreheads. The old man was so surprised at what he saw that he lost his balance and fell out of the hollow tree. Fortunately for him the demons did not see him, as the tree was in the background. So, he picked himself up and crept back into the tree.

While he was sitting there and wondering impatiently when he would be able to get home, he heard the sounds of happy music, and then some of the demons began to sing.

"What are these creatures doing?",said the old man to himself. "I will look out, as it sounds very amusing."

On peeping out, the old man saw that the demon chief himself was actually sitting with his back against the tree in which he had taken refuge, and all the other demons were sitting around, some drinking and some dancing. Food and wine were spread before them on the ground, and the demons were evidently having a great entertainment and enjoying themselves immensely.

It made the old man laugh to see their strange antics.

"How amusing this is!" laughed the old man to himself "I am now quite old, but I have never seen anything so strange in all my life."

He was so interested and excited in watching all that the demons were doing, that he forgot himself and stepped out of the tree and stood looking on.

10. What is the old man's view of the demons?

1. What is the 9th number of the Fibonacci sequence that starts as follows?
   (1, 1, 2, 3, 5, 8, 13.....)

   _____

2. Patrick said that the next two terms in the sequence (1, 3, 5...) were 6 and 10 but that is incorrect. What should the next two terms in the sequence be and why?

3. Fill in the table with the correct values revealing the x intercepts of the following function $f(x)=x^2+2x-3$.

x	f(x)
	0
-2	-3
-1	-4
0	-3
	0
2	5

4. Which of the following is an equation of a parabola with x-intercepts at (-2, 0) and (-3, 0) and a y-intercept at (0, 6)?

   Ⓐ $y=x^2-5x+6$
   Ⓑ $y=x^2+5x-6$
   Ⓒ $y=x^2+5x+6$
   Ⓓ $y=x^2-5x-6$

5. **A certain app charges you $0.50 every time you download a song to your phone. Your dad has placed a cap of $20 per month on your purchases for this particular app. If f(s)=0.5s is the cost of the total number of songs you can download per month, what is the domain of the function?**

Ⓐ 40≤x≤0, where s is an integer
Ⓑ 0≤s≤40, where s is an integer
Ⓒ s≥0, where s is an integer
Ⓓ None of these

Day 3

The Open Boat

None of them knew the color of the sky. Their eyes glanced level and were fastened upon the waves that swept toward them. These waves were of the hue of slate, except for the tops, which were of foaming white, and all of the men knew the colors of the sea. The horizon narrowed and widened and dipped and rose, and at all times its edge was jagged with waves that seemed thrust up in points like rocks.

Many a man ought to have a bathtub larger than the boat which here rode upon the sea. These waves were most wrongfully and barbarously abrupt and tall, and each froth-top was a problem in small boat navigation.

The cook squatted in the bottom and looked with both eyes at the six inches of boat's side which separated him from the ocean. His sleeves were rolled over his fat forearms, and the two flaps of his unbuttoned vest dangled as he bent to bail out the boat. Often he said: "Oh! That was a narrow clip." As he remarked it, he invariably gazed eastward over the broken sea.

The oiler, steering with one of the two oars in the boat, sometimes raised himself suddenly to keep clear of water that swirled in over the stern. It was a thin little oar, and it seemed often ready to snap.

The correspondent, pulling at the other oar, watched the waves and wondered why he was there.

The injured captain, lying in the bow, was at this time buried in that profound dejection and indifference which comes, temporarily at least, to even the bravest and most enduring. When without warning the firm fails, the army loses, and the ship goes down. The mind of the master of a vessel is rooted deep in the timbers of her. Though he commanded for a day or a decade, this captain had on him the stern impression of a scene in the greys of dawn of seven turned faces. Later a stump of a top-mast with a white ball on it that slashed to and fro at the waves went low and lower and down. Thereafter there was something strange in his voice. Although steady, it was deep with mourning and of a quality beyond oration or tears.

"Keep her a little more south, Billie," said he.

"A little more south,'sir," said the oiler in the stern.

A seat in this boat was not unlike a seat upon a bucking bronco and by the same token, a bronco is not much smaller. The craft pranced and reared and plunged like an animal. As each wave came and she rose for it, she seemed like a horse making at a fence outrageously high. The manner of her scramble over these walls of water is a mystic thing, and moreover at the top of them were ordinarily these problems in white water. The foam racing down from the summit of each wave required a new leap and a leap from the air. Then, after scornfully bumping a crest, she would slide and race and splash down a long incline. It would arrive bobbing and nodding in front of the next menace.

A singular disadvantage of the sea lies in the fact that after successfully surmounting one wave, you discover that there is another behind. It is just as important and just as nervously anxious to do something effective in the way of swamping boats. In a ten-foot dinghy, one can get an idea of the resources of the sea in the line of waves that is not probable to the average experience which is never at sea in a dinghy. As each wall of water approached, it shut all else from the view of the men in the boat. It was not difficult to imagine that this particular wave was the final outburst of the ocean and the last effort of the grim water. There was a terrible grace in the move of the waves, and they came in silence, save for the snarling of the crests.

In the pale light, the faces of the men must have been grey. Their eyes must have glinted in strange ways as they gazed steadily astern. Viewed from a balcony, the whole thing would doubtlessly have been weirdly picturesque. However, the men in the boat had no time to see it. If they had leisure, there were other things to occupy their minds. The sun swung steadily up the sky, and they knew it was broad day because the color of the sea changed from slate to emerald-green, streaked with amber lights, and the foam was like tumbling snow. The process of the breaking day was unknown to them. They were aware only of this effect upon the color of the waves that rolled toward them.

Excerpt from A Letter from Joseph Conrad

To this writer of the sea, the sea was not an element. It was a stage, where displayed an exhibition of valor and of such achievement as the world had never seen before. The greatness of that achievement cannot be pronounced imaginary since its reality has affected the destinies of nations. Nevertheless, in its grandeur, it has all the remoteness of an ideal.

6. **Which statement most accurately compares the description of the sea in the letter to the sea in Crane's story?**

   Ⓐ They both reveal a mystery surrounding the sea that a few discovered.
   Ⓑ They both convey the vastness of the sea as an ongoing entity.
   Ⓒ They both reveal the beauty of the sea as a visual image.
   Ⓓ They both depict the sea as always receiving attention from those near or on the sea.

John and Mary Lamb's Version

There was a law in the city of Athens which gave to its citizens the power of compelling their daughters to marry whomsoever they pleased. For upon a daughter's refusing to marry the man her father had chosen to be her husband, the father was empowered by this law to cause her to be put to death. This is because fathers do not often desire the death of their own daughters even though they do happen to prove a little refractory. This law was seldom or never put in execution. Perhaps the young ladies of that city were not infrequently threatened by their parents with the terrors of it.

There was one instance, however, of an old man, whose name was Egeus, who actually did come before Theseus (at that time the reigning duke of Athens), to complain that his daughter Hermia, whom he had commanded to marry Demetrius, a young man of a noble Athenian family, refused to obey him because she loved another young Athenian named Lysander. Egeus demanded justice of Theseus and desired that this cruel law might be put in force against his daughter.

Hermia pleaded in excuse for her disobedience that Demetrius had formerly professed love for her dear friend Helena. Helena loved Demetrius to distraction, but this honorable reason which Hermia gave for not obeying her father's command moved not the stern Egeus.

William Shakespeare's Version

EGEUS: Happy be Theseus, our renowned Duke!
THESEUS: Thanks, good Egeus; what's the news with thee?
EGEUS: Full of vexation come I, with complaint
Against my child, my daughter Hermia.
Stand forth, Demetrius. My noble lord,
This man hath my consent to marry her.
Stand forth, Lysander. And, my gracious Duke,
This man hath bewitch'd the bosom of my child.
Thou, thou, Lysander, thou hast given her rhymes,
And interchang'd love-tokens with my child;
Thou hast by moonlight at her window sung,
With feigning voice, verses of feigning love,
And stol'n the impression of her fantasy
With bracelets of thy hair, rings, gawds, conceits,
Knacks, trifles, nosegays, sweetmeats- messengers
Of strong prevailment in unhardened youth;
With cunning hast thou filch'd my daughter's heart;
Turn'd her obedience, which is due to me,
To stubborn harshness. And, my gracious Duke,
Be it so she will not here before your Grace
Consent to marry with Demetrius,
I beg the ancient privilege of Athens:
As she is mine I may dispose of her;
Which shall be either to this gentleman

Or to her death, according to our law
Immediately provided in that case.
THESEUS: What say you, Hermia? Be advis'd, fair maid.
To you your father should be as a god;
One that compos'd your beauties; yea, and one
To whom you are but as a form in wax,
By him imprinted, and within his power
To leave the figure, or disfigure it.
Demetrius is a worthy gentleman.
HERMIA: So is Lysander.
THESEUS: In himself he is;
But, in this kind, wanting your father's voice,
The other must be held the worthier.
HERMIA: I would my father look'd but with my eyes.
THESEUS: Rather your eyes must with his judgment look.
HERMIA: I do entreat your Grace to pardon me.
I know not by what power I am made bold,
Nor how it may concern my modesty
In such a presence here to plead my thoughts;
But I beseech your Grace that I may know
The worst that may befall me in this case,
If I refuse to wed Demetrius.
THESEUS: Either to die the death, or to abjure
For ever the society of men.
Therefore, fair Hermia, question your desires,

7. **What is emphasized in Shakespeare's version compared to the Lambs' version?**

    Ⓐ The father's reasoning for bringing his daughter to court
    Ⓑ The daughter's explanation of why she does not want to marry Demetrius
    Ⓒ The background of Demetrius
    Ⓓ The background of the law for punishing disobedient daughters

The McWilliamses and the Burglar Alarm

The conversation drifted smoothly and pleasantly along from weather to crops, from crops to literature, from literature to scandal, from scandal to religion. Then it took a random jump and landed on the subject of burglar alarms. Now for the first time Mr. McWilliams showed feeling. Whenever I perceive this sign on this man's dial, I comprehend it, lapse into silence and give him opportunity to unload his heart. He said but with ill-controlled emotion.

"I will not give one single cent for a burglar alarm, Mr. Twain—not a single cent—and I will tell you why. When we were finishing our house, we found we had a little cash left over on account of the plumber not knowing it. I was for enlarging the hearth with it because I was always unaccountably down on the hearth somehow. However, Mrs. McWilliams said, "No, let's have a burglar alarm." I agreed to this compromise. I will explain that whenever I want a thing, and Mrs. McWilliams wants another thing, and we decide upon the thing that Mrs. McWilliams wants—as we always do—she calls that a compromise. The man came up from New York and put in the alarm and charged three hundred and twenty-five dollars for it. He said we could sleep without uneasiness now. So we did for a while—say a month. Then one night we smelled smoke, and I was advised to get up and see what the matter was. I lit a candle and started toward the stairs. I met a burglar coming out of a room with a basket of our tin utensils which he had mistaken for solid silver in the dark. He was smoking a pipe. I said, 'My friend, we do not allow smoking in this room.' He said he was a stranger and could not be expected to know the rules of the house. I said he had been in many houses just as good as this one, and it had never been objected to before. He added that as far as his experience went, such rules had never been considered to apply to burglars, anyway.

"I said, 'Smoke along, then if it is the custom. However, waiving all that, what business have you to be entering this house in this furtive and clandestine way, without ringing the burglar alarm?'

"He looked confused and ashamed and said with embarrassment, 'I beg a thousand pardons. I did not know you had a burglar alarm else I would have rung it. I beg you will not mention it where my parents may hear of it, for they are old and feeble. Such a seemingly cruel breach of the holy conventionalities of our Christian civilization might all too rudely separate the frail bridge. May I trouble you for a match?'

"I said, 'Your sentiments do you honor, but if you will allow me to say it, the metaphor is not your best hold. To return to business, how did you get in here?'

"'Through a second-story window.'

"It was even so. I redeemed the utensils at the pawnbroker's rates, less cost of advertising, bade the burglar goodnight, closed the window after him, and retired to headquarters to report. Next morning, we sent for the burglar-alarm man, and he came up and explained that the reason the alarm did not 'go off' was that no part of the house but the first floor was attached to the alarm. This was simply idiotic because one might as well have no armor on at all in battle as to have it only on his legs. The expert now put the whole second story on the alarm, charged three hundred dollars for it, and went his way. By and by, one night, I found a burglar in the third story about to start down a ladder with a lot of miscellaneous property. My first impulse was to crack his head with a billiard cue, but my second was to refrain from this attention because he was between me and the cue rack. The second impulse was plainly the soundest, so I refrained and proceeded to compromise. I redeemed the property at former rates, after deducting ten percent for use of the ladder. It was my ladder. The next day we sent down for the expert once more and had the third story attached to the alarm for three hundred dollars.

"By this time the burglar alarm had grown to difficult dimensions. It had forty-seven tags on it, marked with the names of the various rooms and chimneys, and it occupied the space of an ordinary wardrobe. The gong was the size of a wash-bowl and was placed above the head of our bed. There was a wire from the house to the coachman's quarters in the stable and a noble gong alongside his pillow.

After spending years of trying to make sure their house is secure, Mr. McWilliams would receive a bill like the one below from an expert itemizing the materials used in securing the house.

Wire	$2.15
Nipple	.75
Two hours of labor	1.50
Wax	.47
Tape	.34
Screws	.15
Recharging battery	.98
Three hours' labor	2.25
String	.02
Lard	.66
Pond's Extract	1.25
Springs at 50	2.00
Railroad fares	7.25

8. What is the list of items that Mr. McWilliams had to pay for?

9. What is the reason for including the list of items that Mr. McWilliams had to pay an expert to fix?

Ⓐ To allow the reader to sympathize with all the expenses the McWilliamses are being charged
Ⓑ To show how a burglar alarm takes a lot of parts to make it work
Ⓒ To display how difficult it is to make a house burglar-proof
Ⓓ To emphasize how technical a burglar alarm is

Havells Stories from the Odyssey

Thence they came to the land of the Cyclopes, a rude and monstrous tribe but favored of the immortal gods by whose bounty they live. They toil not, neither do they sow, nor till the ground, but the earth of herself brings forth for them a bountiful living of wheat and barley and huge swelling clusters of the grape. Naught know they of law or civil life, but each lives in his cave on the wild mountainside, dwelling apart, careless of his neighbors with his wife and children.

It was a dark, cloudy night, and a thick mist overspread the sea when suddenly Odysseus heard the booming of breakers on a rocky shore. Before an order could be given, or any measure taken for the safety of the ships, the little fleet was caught by a strong landward current and whirled recklessly through a narrow passage between the cliffs into a land-locked harbor. Drawing their breath with relief at their wonderful escape, they beached their vessels on the level sand and lay down to wait for the day.

In the morning they found that they had been driven to the landward shore of a long island which formed a natural breakwater to a spacious bay with a narrow entrance at either end. The island was thickly covered with woods giving shelter to a multitude of wild goats, its only inhabitants. For the Cyclopes have no ships, so that the goats were left in undisturbed possession, though the place was well suited for human habitation with a deep, rich soil and plentiful springs of water.

The first care of Odysseus was to supply the crews of his vessels, which were twelve in number, with fresh meat. Armed with bows and spears, he and a picked body of men scoured the woods in search of game. They soon obtained a plentiful booty, and nine goats were assigned to each vessel with ten for that of Odysseus. So, all that day till the setting of the sun they sat and feasted on fat venison and drank of the wine which they had taken in their raid on the Thracians.

Early the next morning Odysseus manned his own galley and set forth to explore the mainland leaving the rest of the crews to await his return on the island. As they drew near the opposite shore of the bay, the mariners came in view of a gigantic cavern overshadowed by laurel-trees. Round the front of the cavern was a wide courtyard rudely fenced with huge blocks of stone and unhewn trunks of trees.

Homer's "Odyssey"

"The land of Cyclops first, a savage kind,
Nor tamed by manners, nor by laws confined:
Untaught to plant, to turn the glebe, and sow,
They all their products to free nature owe:
The soil, untill'd, a ready harvest yields,
With wheat and barley wave the golden fields;
Spontaneous wines from weighty clusters pour,
And Jove descends in each prolific shower,
By these no statues and no rights are known,
No council held, no monarch fills the throne;
But high on hills, or airy cliffs, they dwell,
Or deep in caves whose entrance leads to hell.
Each rules his race, his neighbour not his care,

Heedless of others, to his own severe.

10. Even though each version of the story provides different details, they both include which idea?

Ⓐ Odysseus's thoughts about the Cyclops
Ⓑ the Cyclops's economy
Ⓒ the layout of the land
Ⓓ the anger of the Cyclops

Challenge Yourself!

- **Interpreting Functions**
- **Analyze the Representation**

https://www.lumoslearning.com/a/dc9-8

See Page 7 for Signup details

Day 3

1. Jacob read a 126 page book in 4 days. At what rate was he reading?

2. Damon drove 120 miles in 2 hours and Ashley drove 135 miles in 2.3 hours. Ashley insists that she was traveling at a faster rate than Damon. Is she correct, why?

3. Fill in the blank with the correct values for the Quadratic parent function.

x	f(x)
-2	4
-1	1
0	0
1	1
2	

4. Which function is equivalent to $g(x) = x^4 + x^8 - x^{10}$?

Ⓐ $g(x) = x^4 (x + x^5 - x^6)$
Ⓑ $g(x) = x^4 (1 + x^4 - x^6)$
Ⓒ $g(x) = -x^4 (x^6 + x^4 - 1)$
Ⓓ $g(x) = x^2 (x^2 + x^4 - x^6)$

5. **Which graph could represent a Linear function?**

 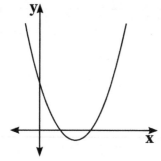

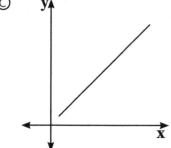

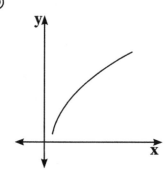

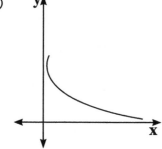

Analyze How An Author Draws

John and Mary Lamb's Version

There was a law in the city of Athens which gave to its citizens the power of compelling their daughters to marry whomsoever they pleased. For upon a daughter's refusing to marry the man her father had chosen to be her husband, the father was empowered by this law to cause her to be put to death. This is because fathers do not often desire the death of their own daughters even though they do happen to prove a little refractory. This law was seldom or never put in execution. Perhaps the young ladies of that city were not infrequently threatened by their parents with the terrors of it.

There was one instance, however, of an old man, whose name was Egeus, who actually did come before Theseus (at that time the reigning duke of Athens), to complain that his daughter Hermia, whom he had commanded to marry Demetrius, a young man of a noble Athenian family, refused to obey him because she loved another young Athenian named Lysander. Egeus demanded justice of Theseus and desired that this cruel law might be put in force against his daughter.

Hermia pleaded in excuse for her disobedience that Demetrius had formerly professed love for her dear friend Helena. Helena loved Demetrius to distraction, but this honorable reason which Hermia gave for not obeying her father's command moved not the stern Egeus.

William Shakespeare's Version

EGEUS: Happy be Theseus, our renowned Duke!
THESEUS: Thanks, good Egeus; what's the news with thee?
EGEUS: Full of vexation come I, with complaint
Against my child, my daughter Hermia.
Stand forth, Demetrius. My noble lord,
This man hath my consent to marry her.
Stand forth, Lysander. And, my gracious Duke,
This man hath bewitch'd the bosom of my child.
Thou, thou, Lysander, thou hast given her rhymes,
And interchang'd love-tokens with my child;
Thou hast by moonlight at her window sung,
With feigning voice, verses of feigning love,
And stol'n the impression of her fantasy
With bracelets of thy hair, rings, gawds, conceits,
Knacks, trifles, nosegays, sweetmeats- messengers
Of strong prevailment in unhardened youth;
With cunning hast thou filch'd my daughter's heart;
Turn'd her obedience, which is due to me,
To stubborn harshness. And, my gracious Duke,
Be it so she will not here before your Grace
Consent to marry with Demetrius,

I beg the ancient privilege of Athens:
As she is mine I may dispose of her;
Which shall be either to this gentleman

Or to her death, according to our law
Immediately provided in that case.
THESEUS: What say you, Hermia? Be advis'd, fair maid.
To you your father should be as a god;
One that compos'd your beauties; yea, and one
To whom you are but as a form in wax,
By him imprinted, and within his power
To leave the figure, or disfigure it.
Demetrius is a worthy gentleman.
HERMIA: So is Lysander.
THESEUS: In himself he is;
But, in this kind, wanting your father's voice,
The other must be held the worthier.
HERMIA: I would my father look'd but with my eyes.
THESEUS: Rather your eyes must with his judgment look.
HERMIA: I do entreat your Grace to pardon me.
I know not by what power I am made bold,
Nor how it may concern my modesty
In such a presence here to plead my thoughts;
But I beseech your Grace that I may know
The worst that may befall me in this case,
If I refuse to wed Demetrius.
THESEUS: Either to die the death, or to abjure
For ever the society of men.
Therefore, fair Hermia, question your desires,

6. **Read the lines below from Shakespeare's version and John Mary Lamb's version. In Shakespeare's version, the father, Egeus, is referred to as a god, but how is the father described in John and Mary Lamb's version?**

Shakespeare:
What say you, Hermia? Be advis'd, fair maid.
To you your father should be as a god;

John and Mary Lamb:
but this honorable reason which Hermia gave for not obeying her father's command moved not the stern Egeus.

Ⓐ The same way as in Shakespeare's version
Ⓑ As a concerned father
Ⓒ As the complete opposite
Ⓓ With compassion towards his daughter

7. **How did the Lambs relay the information about the law for punishing their daughters for refusing to marry the father's choice for them?**

Ⓐ By using a similar format as Shakespeare
Ⓑ By providing it as background information
Ⓒ By considering it to be part of a student's prior knowledge
Ⓓ As part of the plot

Havells Stories from the Odyssey

Thence they came to the land of the Cyclopes, a rude and monstrous tribe but favored of the immortal gods by whose bounty they live. They toil not, neither do they sow, nor till the ground, but the earth of herself brings forth for them a bountiful living of wheat and barley and huge swelling clusters of the grape. Naught know they of law or civil life, but each lives in his cave on the wild mountainside, dwelling apart, careless of his neighbors with his wife and children.

It was a dark, cloudy night, and a thick mist overspread the sea when suddenly Odysseus heard the booming of breakers on a rocky shore. Before an order could be given, or any measure taken for the safety of the ships, the little fleet was caught by a strong landward current and whirled recklessly through a narrow passage between the cliffs into a land-locked harbor. Drawing their breath with relief at their wonderful escape, they beached their vessels on the level sand and lay down to wait for the day.

In the morning they found that they had been driven to the landward shore of a long island which formed a natural breakwater to a spacious bay with a narrow entrance at either end. The island was thickly covered with woods giving shelter to a multitude of wild goats, its only inhabitants. For the Cyclopes have no ships, so that the goats were left in undisturbed possession, though the place was well suited for human habitation with a deep, rich soil and plentiful springs of water.

The first care of Odysseus was to supply the crews of his vessels, which were twelve in number, with fresh meat. Armed with bows and spears, he and a picked body of men scoured the woods in search of game. They soon obtained a plentiful booty, and nine goats were assigned to each vessel with ten for that of Odysseus. So, all that day till the setting of the sun they sat and feasted on fat venison and drank of the wine which they had taken in their raid on the Thracians.

Early the next morning Odysseus manned his own galley and set forth to explore the mainland leaving the rest of the crews to await his return on the island. As they drew near the opposite shore of the bay, the mariners came in view of a gigantic cavern overshadowed by laurel-trees. Round the front of the cavern was a wide courtyard rudely fenced with huge blocks of stone and unhewn trunks of trees.

Homer's "Odyssey"

"The land of Cyclops first, a savage kind,
Nor tamed by manners, nor by laws confined:
Untaught to plant, to turn the glebe, and sow,
They all their products to free nature owe:
The soil, untill'd, a ready harvest yields,
With wheat and barley wave the golden fields;
Spontaneous wines from weighty clusters pour,

And Jove descends in each prolific shower,
By these no statues and no rights are known,
No council held, no monarch fills the throne;
But high on hills, or airy cliffs, they dwell,
Or deep in caves whose entrance leads to hell.
Each rules his race, his neighbour not his care,
Heedless of others, to his own severe.

8. Match the following details from each selection to the version where they are found.

x	Havell's version	Homer's original version
Created a sense of fear when Odysseus arrived	◯	◯
Portrays the Cyclops as a destructive force	◯	◯
Described the lack of government where the Cyclops lived	◯	◯

9. What difference is found in Homer's version showing that the Cyclops were productive with their agriculture?

Ⓐ A few different pieces of equipment are used in the agricultural process.
Ⓑ The Cyclops had big feasts and prepared the foods for these feasts.
Ⓒ Wine was made from the clusters of grapes grown.
Ⓓ Cyclops grew a variety of crops on their land.

10. In the passage, Cyclops did not interact with those around them.
What details from the original version supports this claim?

Challenge Yourself!

- **Interpreting Functions**
- **Analyze How An Author Draws**

https://www.lumoslearning.com/a/dc9-9

Day 4

See Page 7 for Signup details

Day 5

1. Two functions are represented in different ways: The table shows values of f(x), which is a quadratic function.

x	f(x)
2	2
4	8
6	18
8	32

The graph shows the values of g(x), which is a linear function.

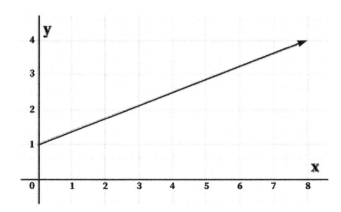

Which function grows at a faster rate when x=6?

Ⓐ f(x)
Ⓑ They grow at the same rate
Ⓒ g(x)
Ⓓ It's impossible to tell

2. Two functions are represented in different ways:
   The table shows values of h(x), which is a quadratic function.

x	h(x)
3	1
6	4
9	9
12	16

The function k(x), which is a linear function, is k(x)=2x−7. Which function grows at a faster rate between x=3 and x=9?

Ⓐ h(x)
Ⓑ They grow at the same rate
Ⓒ k(x)
Ⓓ It's impossible to tell

3. One dollar is invested in an account that accumulates 20% interest every year. What type of function would represent this situation?

Ⓐ linear
Ⓑ quadratic
Ⓒ exponential growth
Ⓓ exponential decay

4. The temperature began at 2°C and decreases 5°C each hour. What type of function would represent this situation?

Ⓐ linear
Ⓑ quadratic
Ⓒ exponential growth
Ⓓ exponential decay

5. The pressure in a chamber of a compressed oxygen tank begins at -2 PSI (pounds per square inch) and every day, the pressure increases 20% more than the previous day. What type of function would represent this situation?

Ⓐ linear
Ⓑ quadratic
Ⓒ exponential growth
Ⓓ exponential decay

6. When a piece of literature is reset in a different time and place, this is called a(n):

Ⓐ Transformation
Ⓑ Adaptation
Ⓒ Sequel
Ⓓ Drama

The Three Little Pigs
by Fran Havard

It was a time of space travel, and the three little pigs were getting ready to take off. They knew they had to have the best shuttle around. There were many enemies in this distant land, and rumor has it some had very big teeth and lots of hair. The first little pig decided to build his space shuttle out of fiberglass because it was cheap and readily available. Unfortunately for him, the enemies knew exactly how to destroy that shuttle and it took one huff and puff to send that little pig crashing back to earth.

7. In the excerpt, what elements have been adapted from the classic fairy tale "The Three Little Pigs"?

Ⓐ Character
Ⓑ Theme
Ⓒ Antagonist
Ⓓ All of the Above

8. In the excerpt, the writer uses what words to highlight the wolf in this adapted tale?

9. In the original tale of The Three Little Pigs, the writer uses hay to build the first house. How is it different in this futuristic version?

10. In this version of The Three Little Pigs, the crucial difference in the adaptation is in the _____.

Ⓐ Setting
Ⓑ Character
Ⓒ Theme
Ⓓ Symbolism
Ⓔ None of the Above

Challenge Yourself!

- **Interpreting Functions**
- **Analyze How An Author Draws**

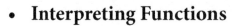

https://www.lumoslearning.com/a/dc9-10

Day 5

See Page 7 for Signup details

Bake Your Way to Summer Fun!

When you're looking for ways to have fun this summer look no further than your kitchen. Baking is a delicious way to be creative and learn a useful skill at the same time. Did you know that baking has been around for literally centuries? Cake dates back to the 13th century according to foodtimeline. org. It's evolved from a honey or fruit-sweetened bread, to the decorated layered masterpieces in a multitude of flavors that we have today. If you've never tried baking before this blog post will walk you through the steps of what you'll need to get started. Before you know it, you'll be sharing your sweet new hobby with all of your friends.

Step 1

You'll need a recipe. Use Google and search for a simple cake recipe for beginners online. I was able to find this easy vanilla cake recipe on Food Network's Website. It has easy-to-follow directions which are perfect for beginners and simple ingredients that you probably already have on hand in your kitchen. Read over the entire recipe once or twice to familiarize yourself with all the steps needed to complete your cake.
Vanilla cake recipe: https://www.foodnetwork.com/recipes/food-network-kitchen/basic-vanilla-cake-recipe-2043654

Step 2

Gather all the ingredients your recipe calls for next. Before you begin to measure or mix anything, you need to make sure you have everything your recipe needs. The amount needed will be written beside each ingredient. An example is 2 C sugar, which means you will need 2 cups of sugar. Measurements are abbreviated on recipes. C stands for cups. Tsp stands for teaspoon. Tbsp stands for tablespoon. Sugar, oil, eggs, flour, salt, some type of flavoring, and baking powder or baking soda are the usual items that go together to make a cake.

Step 3

Next, gather all of your baking tools. You will need a variety of measuring cups and spoons for dry and liquid ingredients. You'll also need a large bowl, a spoon, whisk, or handheld mixer. A stand mixer can be used instead of a hand-held mixer if you have one. It's also a good idea to use a spatula to scrape the sides of the bowl while you're mixing the ingredients. You can scrape the sides and bottom of the bowl making sure everything is mixed thoroughly. The last item you'll need is a pan to bake the cake in. Each recipe will specify the size pan needed. Most pans have their measurements stamped on the bottom so you don't have to guess the size.

Step 4

Now for the fun part — mixing the ingredients! Follow the directions on your recipe exactly. You will need to combine the ingredients in the order that your recipe states. Baking is science and certain ingredients interact with others to create specific outcomes such as a moist cake or a light and fluffy cake. This is why all ingredients need to be combined in a certain order and measured precisely. Follow the directions!

Step 5

Bake your cake and get ready for your kitchen to be filled with wonderful aromas! Make sure to pre-heat your oven and use a timer. To check if your cake is done, insert a toothpick or a knife into the center of the cake. If the toothpick or knife comes out clean then your cake is done. Pull the baked cake out of the oven and set it on the stovetop or a trivet and allow it to cool completely.

Step 6

Frost with icing or enjoy it plain.

Week 2 - PSAT/NMSQT Prep

- **Math**
- **Evidence Based Reading**

https://www.lumoslearning.com/a/slh9-10

See Page 7 for Signup details

Weekly Fun Summer Photo Contest

📷 Take a picture of your summer fun activity and share it on Twitter or Instagram

Use the **#SummerLearning** mention

@LumosLearning on Twitter 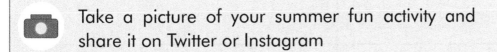 or

@lumos.learning on Instagram 📷

<< Tag friends and increase your chances of winning the contest

Participate and stand a chance to WIN $50 Amazon gift card!

1. **Which function is graphed below?**

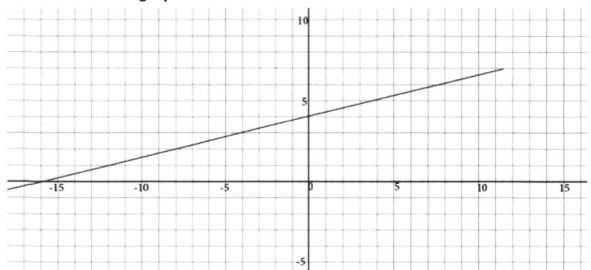

Ⓐ $f(x)=\dfrac{x}{4}+4$

Ⓑ $f(y)=\dfrac{1}{4}y+4$

Ⓒ $f(x)=4x+4$

Ⓓ $f(x)=4x+\dfrac{1}{4}$

2. **Which arithmetic sequence rule produces the term values shown in the table below?**

n	1	2	3	4
a	5	10	15	20

Ⓐ $a_n=5n$

Ⓑ $a_n=10n-5$

Ⓒ $a_n=n+4$

Ⓓ $a_n=4n+3$

3. The graph of an exponential function and a table of values of a linear function are shown below. In which interval do the values in the exponential function begin to exceed the values in the linear function?

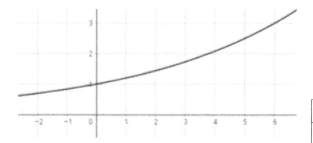

x	0	2	4	6
y	2	2.3	2.6	2.9

Ⓐ [4,5]
Ⓑ [6,7]
Ⓒ [5,6]
Ⓓ [3,4]

4. The graph of a quadratic function and a table of values of a linear function are shown below. In which interval do the values of the linear function begin to exceed the values of the quadratic function?

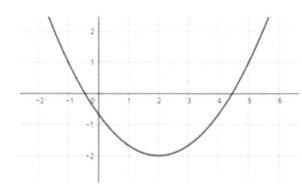

x	0	2	4	6
y	-2	-1	0	1

Ⓐ [4,5]
Ⓑ [6,7]
Ⓒ [5,6]
Ⓓ [3,4]

5. Unless otherwise noted, which base does a logarithm have?

Ⓐ 10
Ⓑ 100
Ⓒ 1000
Ⓓ Not enough information given

Day 1

My Life and Work

I believe that this was the first moving line ever installed. The idea came in a general way from the overhead trolley that the Chicago packers use in packaging beef. We had previously assembled the fly-wheel magneto in the usual method. With one workman doing a complete job, he could turn out from thirty-five to forty pieces in a nine-hour day or about twenty minutes to an assembly. What he did alone was then spread into twenty-nine operations that cut down the assembly time to thirteen minutes, ten seconds. Then we raised the height of the line eight inches which was in 1914 and cut the time to seven minutes. Further experimenting with the speed that the work should move quickly and cut the time down to five minutes. In short, the result is the aid of scientific study one man is now able to do somewhat more than four did only comparatively a few years ago. That line established the efficiency of the method, and we now use it everywhere. The assembling of the motor, formerly done by one man, is now divided into eighty-four operations. Those men do the work that three times their number formerly did. In a short time, we tried out the plan on the framework on the car or the chassis.

About the best we had done in stationary chassis assembling was an average of twelve hours and twenty-eight minutes per chassis. We tried the experiment of drawing the chassis with a rope and windlass down a line two hundred fifty feet long. Six assemblers traveled with the chassis and picked up the parts from piles placed along the line. This rough experiment reduced the time to five hours fifty minutes per chassis. In the early part of 1914, we elevated the assembly line. We had adopted the policy of "man-high" work. We had one line twenty-six and three-quarter inches and another twenty-four and one-half inches from the floor to suit squads of different heights. The waist-high arrangement and a further subdivision of work so that each man had fewer movements cut down the labor time per chassis to one hour thirty-three minutes. Only the chassis was then assembled in the line. The body was placed on "John R. Street" which is the famous street that runs through our Highland Park factories. Now the line assembles the whole car.

It must not be imagined, however, that all this worked out as quickly as it sounds. The speed of the moving work had to be carefully tried out. In the fly-wheel magneto, we first had a speed of sixty inches per minute. That was too fast. Then we tried eighteen inches per minute. That was too slow. Finally, we settled on forty-four inches per minute. The idea is that a man must not be hurried in his work. He must have every second necessary but not a single unnecessary second. We have worked out speeds for each assembly, for the success of the chassis assembly caused us gradually to overhaul our entire method of manufacturing and to put all assembling in mechanically driven lines. The chassis assembling line, for instance, goes at a pace of six feet per minute; the front axle assembly line goes at one hundred eighty-nine inches per minute. In the chassis assembling are forty-five separate operations or stations. The first men fasten four mud-guard brackets to the chassis frame. Then the motor arrives on the tenth operation and so on in detail. Some men do only one or two small operations whereas others do more. The man who places a part does not fasten it because the part may not be fully in place until after several operations later. The man who puts in a bolt does not put on the nut. The man who puts on the nut does not tighten it. On operation number thirty-four, the budding motor gets its gasoline. It has previously received lubrication. On operation number forty-four, the radiator is filled with water, and on operation number forty-five the car drives out onto John R. Street.

6. **Which evidence from the passage supports the fact that the assembly line was created based on another form of an assembly line method?**

   Ⓐ "We had previously assembled the fly-wheel magneto in the usual method."
   Ⓑ "The idea came in a general way from the overhead trolley that the Chicago packers use in packaging beef."
   Ⓒ "I believe that this was the first moving line ever installed."
   Ⓓ "Further experimenting with the speed that the work should move quickly and cut the time down to five minutes."

7. **Which evidence shows that the assembly line has become popular with companies?**

   Ⓐ "About the best we had done in stationary chassis assembling was an average of twelve hours and twenty-eight minutes per chassis."
   Ⓑ "Those men do the work that three times their number formerly did. In a short time, we tried out the plan on the framework on the car or the chassis."
   Ⓒ "The assembling of the motor, formerly done by one man, is now divided into eighty-four operations."
   Ⓓ "That line established the efficiency of the method, and we now use it everywhere."

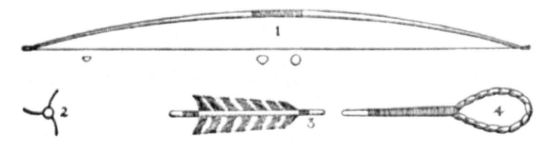

How to Make a Bow and Arrow

In early days, the Indian did not have the modern hunting rifle and was compelled to use bow and arrow in the chase. It is well for the modern boy to understand this weapon, for it can be made with ease and much good fun can be had with it. The Indian bow was short because though less efficient it was easier to carry than a long one, yet it did not lack power. We hear that many times Indians shot so hard that their arrows appeared on the far side of the animal, but the long bow, such was used by the old English archers or bowmen, was much the more powerful.

To make the bow, take a perfectly sound, straight, well-seasoned stick of about your own height and mark off a space as wide as your hand in the middle for a handle. This space should be left round and about an inch thick. The balance of the stick should be shaved down flat on one side for the front and rounded on the other for the back until it is about one inch wide and three-fourths of an inch thick near the handle tapering to about half that at the ends which are then notched for the cord. Next put on the cord and bend it until it is about five inches from the bow at the center. If one end bends more than the other, shave the other end until it becomes even.

After it is trimmed down to your strength, finish it with sand-paper. The best woods to use are apple, black walnut, slippery elm, mountain ash, or hickory.

More difficult to make than the bow is the arrow. The Indians made arrows of reeds and of straight shoots of arrow-wood or of elder, but we make better arrows out of hickory or ash. The arrow should be twenty-five inches long, round, and three-eighths of an inch thick. It should be notched at one end to take the bow-string, and just below this notch, it should have three feathers set around at equal intervals. These feathers are best made from either turkey or goose wings. With a sharp knife, cut a strip of the midrib on which is the vein; make three pieces, each two to three inches long. The Indians used to leave the midrib projecting at each end, and by these lash the feathers to the arrow without gluing, but it is easier to glue them and the arrows fly better. Indian arrow-heads were made of sharp flints or pieces of stone lashed to the arrowheads, but the best way to make them now is like the ferrule of an umbrella as this keeps the shaft from splitting. After this the arrow should be painted, both to keep it from warping and to make it easier to find in the forest by its bright color.

After the bow and arrows are made, one must learn to shoot with them. It is good to begin with the target close at hand, and gradually increase the distance as the archer becomes more expert. The Indians generally used their bows at short range so that it was easy to hit the mark, and considered rapid firing more important. In their competitions, the prize was given to him who should have the most arrows in the air at once, and it has been said that their record was eight.

8. Match the evidence with the information it supports.

	"The strength of the Indian's bow and arrow is shown compared to that of others."	"Supplies are specific for optimum results."	"Techniques are described to keep the arrows strong."
After this the arrow should be painted, both to keep it from warping and to make it easier to find in the forest by its bright color.	○	○	○
We hear that many times Indians shot so hard that their arrows appeared on the far side of the animal, but the long bow, such was used by the old English archers or bowmen, was much the more powerful.	○	○	○
These feathers are best made from either turkey or goose wings.	○	○	○

How It Works

By attaching a small wheel to the end of a Morse-sounder lever, by arranging an ink-well for the wheel to dip into when the end falls, and by moving a paper ribbon slowly along for the wheel to press against when it rises, a self-recording Morse inker is produced. The ribbon-feeding apparatus is set in motion automatically by the current and continues to pull the ribbon along until the message is completed.

The Hughes type-printer covers a sheet of paper with printed characters in bold Roman type. The transmitter has a keyboard on which are marked letters, signs, and numbers. Also, a type-wheel, with the characters on its circumference, rotates by electricity. The receiver contains mechanisms for rotating another type-wheel synchronously—that is, in time—with the first, for shifting the wheel across the paper, for pressing the paper against the wheel, and for moving the paper when a fresh line is needed. These are too complicated to be described here in detail. By means of relays, one transmitter may be made to work five hundred receivers. In London a single operator controlling a keyboard in the central dispatching office causes typewritten messages to spell themselves out simultaneously in machines distributed all over the metropolis.

The tape machine resembles what was just described in many details. The main difference is that it prints on a continuous ribbon instead of on sheets.

Automatic electric printers of some kind or others are to be found in the foyers of all the principal hotels and clubs of our large cities and in the offices of bankers, stockbrokers, and newspaper editors. In London alone, over 500 million words are printed by the receivers in a year.

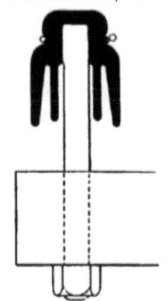

Fig. - Section of a telegraph wire insulator on its arm. The shaded circle is the line wire, the two blank circles indicate the wire which ties the line wire to the insulator.

9. **What evidence reveals that the receiver serves a large number of functions?**

   Ⓐ "The tape machine resembles what was just described in many details."
   Ⓑ "These are too complicated to be described here in detail."
   Ⓒ "By means of relays, one transmitter may be made to work five hundred receivers."
   Ⓓ "The receiver contains mechanisms for rotating another type-wheel synchronously."

Whales, Dolphins, and Porpoises of the Western North Atlantic

All five species of large whales with a dorsal fin belong to the same major baleen whale group, the balaenopterid whales or rorquals. All are characterized by the presence of a series of ventral grooves usually visible on stranded specimens and the length and number of which are diagnostic to species. In addition, all species, with the exception of the humpback whale, have at least one distinctive (though often not prominent) ridge along the head from just in front of the blowhole to near the tip of the snout. (The humpback whale, on the other hand, is distinguished by numerous knobs, some of which are located along the line of the head ridge with others scattered on the top of the head.) In Bryde's whale, the single head ridge characteristic of the other rorquals is supplemented by two auxiliary ridges one on each side of the main ridge.

At sea, these whales often appear very similar and must be examined carefully before they can be reliably identified.

In general, though the characteristics of behavior may vary from one encounter to the next, based on the activities in which the animal is engaged. Whales in this group may be distinguished from each other on the basis of differences in 1) the size, shape, and position of the dorsal fin and the timing of its appearance on the surface relative to the animal's blow (in general, the larger the whale, the smaller the dorsal fin—the further back its position and the later its appearance on the surface after the animal's blow); 2) the height of the body in the area of the dorsal fin, relative to the size of the dorsal fin, which is exposed as the animal sounds; 3) sometimes the blow rate and movement patterns; and 4) the shape and color of the head.

Despite variability in behavior by members of the same species from one encounter to the next, an observer can greatly increase the reliability of his identification by forming the habit of working systematically through a set of characteristics for the species rather than depending on any single characteristic.

There are three species of large whales without a dorsal fin in the western North Atlantic Ocean. Two of these, the bowhead or Greenland whale, and its more widely distributed close relative, the right whale, are baleen whales. The third, the sperm whale, is a toothed whale. The first two have relatively smooth backs without even a trace of a dorsal fin. The sperm whale has a humplike low, thick, dorsal ridge, which from certain views particularly when the animal is humping up to begin a dive may be clearly visible and looks like a fin. But because the profile of that hump and the knuckles which follow it are often not very prominent in this species, it has been classified with the finless big whales.

All three species are characterized by very distinctive blows or spouts. In both the bowhead and the right whales, the projection of the blow upward from two widely separated blowholes assumes a very wide V-shape with two distinct columns, which may be seen when the animals are viewed from front or back. Though this character may be visible under ideal conditions in many of the other baleen whales species as well, it is exaggerated and uniformly distinct in the bowhead and right whales and may be used as one of the primary key characters. In the sperm whale, the blow emanates from a blowhole which is displaced to the left of the head near the front and projects obliquely forward to the animal's left. This blow seen under ideal conditions positively labels a large whale as a sperm whale.

10. **What sentence from the passage supports that the sperm whale is different from the other whales?**

Ⓐ "The third, the sperm whale, is a toothed whale…"

Ⓑ "There are three species of large whales without a dorsal fin in the western North Atlantic Ocean…"

Ⓒ "All three species are characterized by very distinctive blows or spouts…"

Ⓓ "But because the profile of that hump and the knuckles which follow it are often not very prominent in this species, it has been classified with the finless big whales…"

Challenge Yourself!

- **Interpreting Functions**
- **Textual Evidence to Support Analysis**

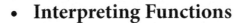

https://www.lumoslearning.com/a/dc9-11

Day 1

See Page 7 for Signup details

Congruence

1. Suppose there are 1,090 hamburger patties in the walk-in cooler at a restaurant at the beginning of the week. The restaurant expects to sell 150 hamburger patties each day. Which function H(p) represents the number of hamburger patties that will be in the cooler after selling them for p days?

   Ⓐ H(p)=1090−150p
   Ⓑ H(p)=−1090+150p
   Ⓒ H(p)=1090+150p
   Ⓓ H(p)=1090p−150

2. Suppose a construction company is preparing to build a large apartment complex and is beginning to receive truckloads of sheets of plywood for the construction project. The company has 150 sheets on hand and each truckload has 80 sheets. Which function P(t) represents the total number of sheets of plywood the company has if t truckloads of plywood have been delivered?

   Ⓐ P(t)=150−80t
   Ⓑ P(t)=150+80t
   Ⓒ P(t)=−150−80t
   Ⓓ P(t)=−150−80t

3. What is defined as a part of a line that is bounded by two distinct end points?

   Ⓐ Angle
   Ⓑ Circle
   Ⓒ Line segment
   Ⓓ Perpendicular line

4. Which of the following is an accurate example of a ray?

   Ⓐ The beam of light leaving a flashlight into the sky.
   Ⓑ A line of people waiting to check out at a store.
   Ⓒ Cars in a traffic jam.
   Ⓓ The equator of the earth.

5. **What would be the coordinates of point F after a translation of three units to the right and four units down?**

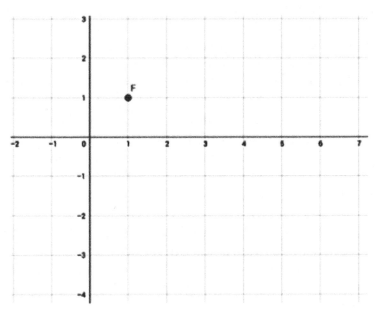

Ⓐ (1, 1)
Ⓑ (1, -3)
Ⓒ (4, 1)
Ⓓ (4, -3)

Determine a Central Idea of a Text

Day 2

Common Sense

Some writers have so been annoyed by society with the government as to leave little or no distinction between them whereas they are not only different but have different origins. Society is produced by our wants and the government by our wickedness. The former promotes our happiness positively by uniting our affections and the latter negatively by restraining our vices. The one encourages communication and the other creates differences. The first is a supporter, and the last is a punisher.

Society in every state is a blessing, but government even at its best state is but a necessary evil. In its worst state, it is an intolerable one. For when we suffer or are exposed to the same miseries by a government, our misfortune is heightened by reflecting that we furnish the means by which we suffer. Government is the badge of lost innocence, and the palaces of kings are built on the ruins of the places of paradise. If the impulses of a clear conscience, clear expectations, and irresistibly obeyed officials are there, then man would need no other lawgiver. However, that is not the case because he finds it necessary to surrender up a part of his property to furnish means for the protection of the rest. This, he is induced to do by the same cautiousness which in every other case advises him out of two evils to choose the least. Wherefore, security should be the true policy from the government, so that it follows whatever form appears most likely to ensure it to us with the least expense and greatest benefit is preferable to all others.

In order to gain a clear and just idea of the design of the government, let us suppose a small number of people settled in some sequestered part of the earth, unconnected with the rest, they will then represent the first peopling of any country or of the world. In this state of natural liberty, society will be their first thought. A thousand reasons will excite them thereto, the strength of one man is so unequal to his wants, and his mind so unfitted for continuous isolation, that he is soon obliged to seek assistance and relief of another. He requires the same. Four or five united people would be able to raise a tolerable dwelling in the midst of a wilderness, but one man might work out of the common period of life without accomplishing anything. When he had chopped for his timber, he could not remove it, nor build with it after it was removed. Therefore, hunger in the meantime would bother him from his work,and every different need calls him in a different way. Disease would disable him from living and reduce him to a state in which he might rather be said to perish than to die.

Thus, necessity would soon form our newly arrived emigrants into society the common blessings of the obligations of law and government unnecessary while they remained perfectly fair to each other. They prevail the first difficulties of emigration which bound them together in a common cause. They will begin to relax in their duty and attachment to each other. This is careless of establishing some form of government to supply the defect of moral virtue.

Some convenient tree will afford them a State-House, under the branches of which, the whole colony may assemble to deliberate on public matters. It is more than probable that their first laws will have the title only of regulations, and be enforced by no other penalty than public disfavor. In this first parliament every man, by natural right, will have a seat.

6. **Match the following statements to either society or government as it is described in the passage.**

	Society	Government
Brings together what we like	◯	◯
Holds in our evil	◯	◯
Created by our desires	◯	◯
Produced by mischief	◯	◯

7. **What is the theme in "Common Sense"? Use evidence from the text to support your ideas.**

My Life and Work

I believe that this was the first moving line ever installed. The idea came in a general way from the overhead trolley that the Chicago packers use in packaging beef. We had previously assembled the fly-wheel magneto in the usual method. With one workman doing a complete job, he could turn out from thirty-five to forty pieces in a nine-hour day or about twenty minutes to an assembly. What he did alone was then spread into twenty-nine operations that cut down the assembly time to thirteen minutes, ten seconds. Then we raised the height of the line eight inches which was in 1914 and cut the time to seven minutes. Further experimenting with the speed that the work should move quickly and cut the time down to five minutes. In short, the result is the aid of scientific study one man is now able to do somewhat more than four did only comparatively a few years ago. That line established the efficiency of the method, and we now use it everywhere. The assembling of the motor, formerly done by one man, is now divided into eighty-four operations. Those men do the work that three times their number formerly did. In a short time, we tried out the plan on the framework on the car or the chassis.

About the best we had done in stationary chassis assembling was an average of twelve hours and twenty-eight minutes per chassis. We tried the experiment of drawing the chassis with a rope and windlass down a line two hundred fifty feet long. Six assemblers traveled with the chassis and picked up the parts from piles placed along the line. This rough experiment reduced the time to five hours fifty minutes per chassis. In the early part of 1914, we elevated the assembly line. We had adopted the policy of "man-high" work. We had one line twenty-six and three-quarter inches and another twenty-four and one-half inches from the floor to suit squads of different heights. The waist-high arrangement and a further subdivision of work so that each man had fewer movements cut down the labor time per chassis to one hour thirty-three minutes. Only the chassis was then assembled in the line. The body was placed on "John R. Street" which is the famous street that runs through our Highland Park factories. Now the line assembles the whole car.

It must not be imagined, however, that all this worked out as quickly as it sounds. The speed of the moving work had to be carefully tried out. In the fly-wheel magneto, we first had a speed of sixty inches per minute. That was too fast. Then we tried eighteen inches per minute. That was too slow. Finally, we settled on forty-four inches per minute. The idea is that a man must not be hurried in his work. He must have every second necessary but not a single unnecessary second. We have worked out speeds for each assembly, for the success of the chassis assembly caused us gradually to overhaul our entire method of manufacturing and to put all assembling in mechanically driven lines. The chassis assembling line, for instance, goes at a pace of six feet per minute; the front axle assembly line goes at one hundred eighty-nine inches per minute. In the chassis assembling are forty-five separate operations or stations. The first men fasten four mud-guard brackets to the chassis frame. Then the motor arrives on the tenth operation and so on in detail. Some men do only one or two small operations whereas others do more. The man who places a part does not fasten it because the part may not be fully in place until after several operations later. The man who puts in a bolt does not put on the nut. The man who puts on the nut does not tighten it. On operation number thirty-four, the budding motor gets its gasoline. It has previously received lubrication. On operation number forty-four, the radiator is filled with water, and on operation number forty-five the car drives out onto John R. Street.

8. Which of the following contributes to the development of the central idea?

Ⓐ the length of time it takes a car to be built using an assembly line
Ⓑ the definition of chassis
Ⓒ when the height of the line was raised
Ⓓ the street where the car was placed

Declaration of Independence

The history of the present King of Great Britain is a history of repeated injuries and seizures, all having in direct object the establishment of an absolute Tyranny over these States. To prove this, let Facts be submitted to a candid world.

He has refused his Assent to Laws, the most wholesome and necessary for the public good.

He has forbidden his Governors to pass Laws of immediate and pressing importance unless suspended in their operation till his Assent should be obtained; and when so suspended, he has utterly neglected to attend to them.

He has refused to pass other Laws for the accommodation of large districts of people, unless those people would relinquish the right of Representation in the Legislature, a right inestimable to them and formidable to tyrants only.

He has called together legislative bodies at places unusual, uncomfortable, and distant from the depository of their Public Records, for the sole purpose of fatiguing them into compliance with his measures.

He has dissolved Representative Houses repeatedly, for opposing with manly firmness his invasions on the rights of the people.

He has refused for a long time, after such dissolutions, to cause others to be elected; whereby the Legislative Powers, incapable of Annihilation, have returned to the People at large for their exercise; the State remaining in the meantime exposed to all the dangers of invasion from without, and convulsions within.

He has endeavored to prevent the population of these States; for that purpose obstructing the Laws of Naturalization of Foreigners; refusing to pass others to encourage their migration hither, and raising the conditions of new Appropriations of Lands.

9. **Match the following statements to either society or government as it is described in the passage.**

	Disagreement of needed regulations	Did not use parts of Parliament bodies	Useless meetings
dissolved Representative Houses repeatedly	◯	◯	◯
refused his Assent to Laws	◯	◯	◯
called together legislative bodies at places unusual, uncomfortable, and distant	◯	◯	◯

How It Works

By attaching a small wheel to the end of a Morse-sounder lever, by arranging an ink-well for the wheel to dip into when the end falls, and by moving a paper ribbon slowly along for the wheel to press against when it rises, a self-recording Morse inker is produced. The ribbon-feeding apparatus is set in motion automatically by the current and continues to pull the ribbon along until the message is completed.

The Hughes type-printer covers a sheet of paper with printed characters in bold Roman type. The transmitter has a keyboard on which are marked letters, signs, and numbers. Also, a type-wheel, with the characters on its circumference, rotates by electricity. The receiver contains mechanisms for rotating another type-wheel synchronously—that is, in time—with the first, for shifting the wheel across the paper, for pressing the paper against the wheel, and for moving the paper when a fresh line is needed. These are too complicated to be described here in detail. By means of relays, one transmitter may be made to work five hundred receivers. In London a single operator controlling a keyboard in the central dispatching office causes typewritten messages to spell themselves out simultaneously in machines distributed all over the metropolis.

The tape machine resembles what was just described in many details. The main difference is that it prints on a continuous ribbon instead of on sheets.

Automatic electric printers of some kind or others are to be found in the foyers of all the principal hotels and clubs of our large cities and in the offices of bankers, stockbrokers, and newspaper editors. In London alone, over 500 million words are printed by the receivers in a year.

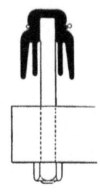

Fig. - Section of a telegraph wire insulator on its arm. The shaded circle is the line wire, the two blank circles indicate the wire which ties the line wire to the insulator.

10. The central idea emerges through the description of the telegraph's _____.

1. **What combination of transformations will map the image ABCD onto A'B'C'D'?**

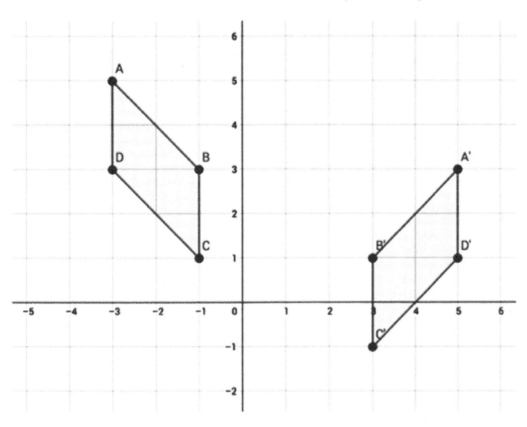

Ⓐ Reflection over the x-axis and translate 2 down
Ⓑ Translate to the right 4 and down 2
● Reflection over the x = 1 and a translation two units down
Ⓓ Reflection over the line y = 1 and translate 4 to the left

2. **What translation rule will map triangle ABC onto triangle A'B'C'?**

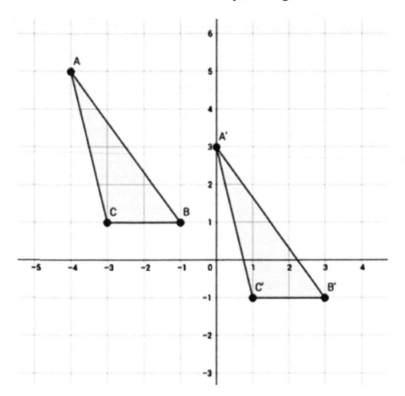

Ⓐ (x + 4, y + 2)
Ⓑ (X + 2, y + 4)
Ⓒ (x - 4, y - 2)
Ⓓ (x + 4, y - 2)

3. **What type of transformation is defined by (x, y) → (x + 1, y – 2) ?**

Ⓐ Reflection
Ⓑ Translation
Ⓒ Rotation
Ⓓ Dilation

4. Which of the following show a translation of three units to the left and two units down of the shape ABCD?

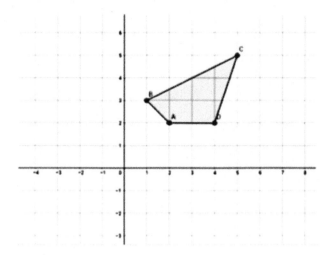

Ⓐ

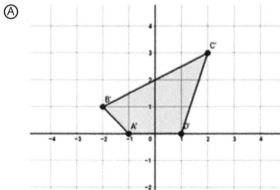

Ⓒ

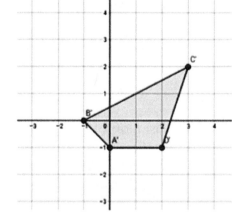

Ⓑ

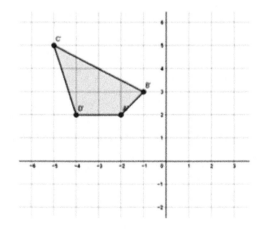

Ⓓ

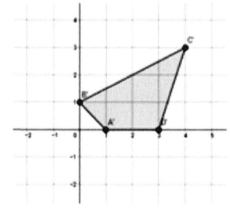

5. **What combination of transformations will map ABCDE onto A'B'C'D'E'?**

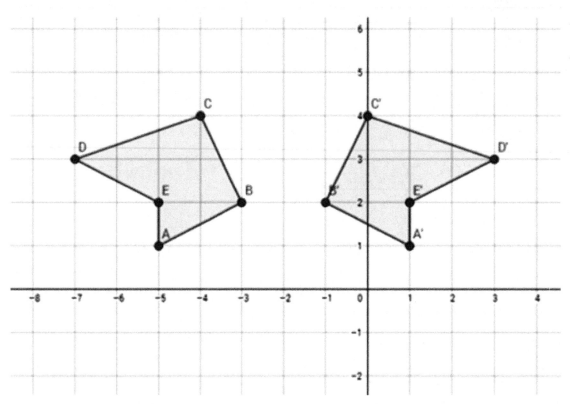

Ⓐ A translation of two units to the right
Ⓑ A reflection across the line y = -2
Ⓒ A reflection across the line x = -2
Ⓓ A 90°clockwise around the point (-2, 2)

Analyze How the Author Unfolds Series of Ideas

Common Sense

Some writers have so been annoyed by society with the government as to leave little or no distinction between them whereas they are not only different but have different origins. Society is produced by our wants and the government by our wickedness. The former promotes our happiness positively by uniting our affections and the latter negatively by restraining our vices. The one encourages communication and the other creates differences. The first is a supporter, and the last is a punisher.

Society in every state is a blessing, but government even at its best state is but a necessary evil. In its worst state, it is an intolerable one. For when we suffer or are exposed to the same miseries by a government, our misfortune is heightened by reflecting that we furnish the means by which we suffer. Government is the badge of lost innocence, and the palaces of kings are built on the ruins of the places of paradise. If the impulses of a clear conscience, clear expectations, and irresistibly obeyed officials are there, then man would need no other lawgiver. However, that is not the case because he finds it necessary to surrender up a part of his property to furnish means for the protection of the rest. This, he is induced to do by the same cautiousness which in every other case advises him out of two evils to choose the least. Wherefore, security should be the true policy from the government, so that it follows whatever form appears most likely to ensure it to us with the least expense and greatest benefit is preferable to all others.

In order to gain a clear and just idea of the design of the government, let us suppose a small number of people settled in some sequestered part of the earth, unconnected with the rest, they will then represent the first peopling of any country or of the world. In this state of natural liberty, society will be their first thought. A thousand reasons will excite them thereto, the strength of one man is so unequal to his wants, and his mind so unfitted for continuous isolation, that he is soon obliged to seek assistance and relief of another. He requires the same. Four or five united people would be able to raise a tolerable dwelling in the midst of a wilderness, but one man might work out of the common period of life without accomplishing anything. When he had chopped for his timber, he could not remove it, nor build with it after it was removed. Therefore, hunger in the meantime would bother him from his work,and every different need calls him in a different way. Disease would disable him from living and reduce him to a state in which he might rather be said to perish than to die.

Thus, necessity would soon form our newly arrived emigrants into society the common blessings of the obligations of law and government unnecessary while they remained perfectly fair to each other. They prevail the first difficulties of emigration which bound them together in a common cause. They will begin to relax in their duty and attachment to each other. This is careless of establishing some form of government to supply the defect of moral virtue.

Some convenient tree will afford them a State-House, under the branches of which, the whole colony may assemble to deliberate on public matters. It is more than probable that their first laws will have the title only of regulations, and be enforced by no other penalty than public disfavor. In this first parliament every man, by natural right, will have a seat.

6. **After reading paragraph 3 of the passage, what do you think the author has included to describe how the government would want to be built by the citizens themselves?**

7. **Describe the connection Thomas Paine makes between the government and their citizens.**

My Life and Work

I believe that this was the first moving line ever installed. The idea came in a general way from the overhead trolley that the Chicago packers use in packaging beef. We had previously assembled the fly-wheel magneto in the usual method. With one workman doing a complete job, he could turn out from thirty-five to forty pieces in a nine-hour day or about twenty minutes to an assembly. What he did alone was then spread into twenty-nine operations that cut down the assembly time to thirteen minutes, ten seconds. Then we raised the height of the line eight inches which was in 1914 and cut the time to seven minutes. Further experimenting with the speed that the work should move quickly and cut the time down to five minutes. In short, the result is the aid of scientific study one man is now able to do somewhat more than four did only comparatively a few years ago. That line established the efficiency of the method, and we now use it everywhere. The assembling of the motor, formerly done by one man, is now divided into eighty-four operations. Those men do the work that three times their number formerly did. In a short time, we tried out the plan on the framework on the car or the chassis.

About the best we had done in stationary chassis assembling was an average of twelve hours and twenty-eight minutes per chassis. We tried the experiment of drawing the chassis with a rope and windlass down a line two hundred fifty feet long. Six assemblers traveled with the chassis and picked up the parts from piles placed along the line. This rough experiment reduced the time to five hours fifty minutes per chassis. In the early part of 1914, we elevated the assembly line. We had adopted the policy of "man-high" work. We had one line twenty-six and three-quarter inches and another twenty-four and one-half inches from the floor to suit squads of different heights. The waist-high arrangement and a further subdivision of work so that each man had fewer movements cut down the labor time per chassis to one hour thirty-three minutes. Only the chassis was then assembled in the line. The body was placed on "John R. Street" which is the famous street that runs through our Highland Park factories. Now the line assembles the whole car.

It must not be imagined, however, that all this worked out as quickly as it sounds. The speed of the moving work had to be carefully tried out. In the fly-wheel magneto, we first had a speed of sixty inches per minute. That was too fast. Then we tried eighteen inches per minute. That was too slow. Finally, we settled on forty-four inches per minute. The idea is that a man must not be hurried in his work. He must have every second necessary but not a single unnecessary second. We have worked out speeds for each assembly, for the success of the chassis assembly caused us gradually to overhaul our entire method of manufacturing and to put all assembling in mechanically driven lines. The chassis assembling line, for instance, goes at a pace of six feet per minute; the front axle assembly line goes at one hundred eighty-nine inches per minute. In the chassis assembling are forty-five separate operations or stations. The first men fasten four mud-guard brackets to the chassis frame. Then the motor arrives on the tenth operation and so on in detail. Some men do only one or two small operations whereas others do more. The man who places a part does not fasten it because the part may not be fully in place until after several operations later. The man who puts in a bolt does not put on the nut. The man who puts on the nut does not tighten it. On operation number thirty-four, the budding motor gets its gasoline. It has previously received lubrication. On operation number forty-four, the radiator is filled with water, and on operation number forty-five the car drives out onto John R. Street.

8. What does the author do to reveal information about his assembly line?

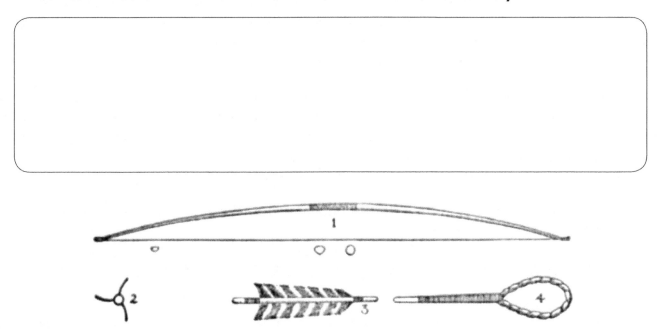

How to Make a Bow and Arrow

In early days, the Indian did not have the modern hunting rifle and was compelled to use bow and arrow in the chase. It is well for the modern boy to understand this weapon, for it can be made with ease and much good fun can be had with it. The Indian bow was short because though less efficient it was easier to carry than a long one, yet it did not lack power. We hear that many times Indians shot so hard that their arrows appeared on the far side of the animal, but the long bow, such was used by the old English archers or bowmen, was much the more powerful.

To make the bow, take a perfectly sound, straight, well-seasoned stick of about your own height and mark off a space as wide as your hand in the middle for a handle. This space should be left round and about an inch thick. The balance of the stick should be shaved down flat on one side for the front and rounded on the other for the back until it is about one inch wide and three-fourths of an inch thick near the handle tapering to about half that at the ends which are then notched for the cord. Next put on the cord and bend it until it is about five inches from the bow at the center. If one end bends more than the other, shave the other end until it becomes even.

After it is trimmed down to your strength, finish it with sand-paper. The best woods to use are apple, black walnut, slippery elm, mountain ash, or hickory.

More difficult to make than the bow is the arrow. The Indians made arrows of reeds and of straight shoots of arrow-wood or of elder, but we make better arrows out of hickory or ash. The arrow should be twenty-five inches long, round, and three-eighths of an inch thick. It should be notched at one end to take the bow-string, and just below this notch, it should have three feathers set around at equal intervals. These feathers are best made from either turkey or goose wings. With a sharp knife, cut a strip of the midrib on which is the vein; make three pieces, each two to three inches long. The Indians used to leave the midrib projecting at each end, and by these lash the feathers to the arrow without gluing, but it is easier to glue them and the arrows fly better. Indian arrow-heads were made of sharp flints or pieces of stone lashed to the arrowheads, but the best way to make them now is like the ferrule of an umbrella as this keeps the shaft from splitting. After this the arrow should be painted, both to keep it from warping and to make it easier to find in the forest by its bright color.

After the bow and arrows are made, one must learn to shoot with them. It is good to begin with the target close at hand, and gradually increase the distance as the archer becomes more expert. The Indians generally used their bows at short range so that it was easy to hit the mark, and considered rapid firing more important. In their competitions, the prize was given to him who should have the most arrows in the air at once, and it has been said that their record was eight.

9. **Which choice best explains how the author describes the process of making a bow and arrow?**

   Ⓐ By including the materials needed and the process of how to make that part of the bow and arrow
   Ⓑ By explaining all of the materials first and then the process
   Ⓒ By describing each part of the bow and arrow before explaining the process for making the bow and arrow
   Ⓓ By explaining why each part of the bow and arrow helps make this the most powerful bow and arrow

10. How are the bow and arrow introduced in this passage?

Congruence (Contd.)

1. Describe the transformation that will carry the shape onto itself. More than one answer may be correct. Choose all that apply.

Ⓐ A reflection over the line m
Ⓑ A rotation 90° clockwise
Ⓒ A rotation 180° clockwise
Ⓓ A rotation 90° counterclockwise

2. Which of the following transformations will preserve the original images lengths and angle measures?

Ⓐ A dilation
Ⓑ A reflection
Ⓒ A vertical stretch
Ⓓ A horizontal stretch

3. Two triangles that are congruent have which of the same measurements that are also equal? More than one answer may be correct. Choose all that apply.

Ⓐ Corresponding side length
Ⓑ Corresponding angle Length
Ⓒ Perimeter
Ⓓ Area

4. **A ladder is leaning up against a wall to reach a window creating a triangle with the building and the ground. If the ladder is moved seven feet to the next window on the same level creating another triangle, such that the distance and angle between the window and the ladder would remain same. Is the new triangle congruent to the original triangle?**

   Ⓐ No they are not congruent since it is a reflection
   Ⓑ No they are not congruent since it is a translation
   Ⓒ Yes they are congruent since it is a reflection
   Ⓓ Yes they are congruent since it is a translation

5. **Which of the following is needed in order to determine if two triangles are congruent?**

   Ⓐ Any two angles that are congruent
   Ⓑ Two angles and a side that are congruent
   Ⓒ Two sides that are congruent and the congruent angle that is in between the two sides
   Ⓓ Two sides that are congruent and an angle that is congruent

Determine the Meaning of Words and Phrases

Day 4

My Life and Work

I believe that this was the first moving line ever installed. The idea came in a general way from the overhead trolley that the Chicago packers use in packaging beef. We had previously assembled the fly-wheel magneto in the usual method. With one workman doing a complete job, he could turn out from thirty-five to forty pieces in a nine-hour day or about twenty minutes to an assembly. What he did alone was then spread into twenty-nine operations that cut down the assembly time to thirteen minutes, ten seconds. Then we raised the height of the line eight inches which was in 1914 and cut the time to seven minutes. Further experimenting with the speed that the work should move quickly and cut the time down to five minutes. In short, the result is the aid of scientific study one man is now able to do somewhat more than four did only comparatively a few years ago. That line established the efficiency of the method, and we now use it everywhere. The assembling of the motor, formerly done by one man, is now divided into eighty-four operations. Those men do the work that three times their number formerly did. In a short time, we tried out the plan on the framework on the car or the chassis.

About the best we had done in stationary chassis assembling was an average of twelve hours and twenty-eight minutes per chassis. We tried the experiment of drawing the chassis with a rope and windlass down a line two hundred fifty feet long. Six assemblers traveled with the chassis and picked up the parts from piles placed along the line. This rough experiment reduced the time to five hours fifty minutes per chassis. In the early part of 1914, we elevated the assembly line. We had adopted the policy of "man-high" work. We had one line twenty-six and three-quarter inches and another twenty-four and one-half inches from the floor to suit squads of different heights. The waist-high arrangement and a further subdivision of work so that each man had fewer movements cut down the labor time per chassis to one hour thirty-three minutes. Only the chassis was then assembled in the line. The body was placed on "John R. Street" which is the famous street that runs through our High-land Park factories. Now the line assembles the whole car.

It must not be imagined, however, that all this worked out as quickly as it sounds. The speed of the moving work had to be carefully tried out. In the fly-wheel magneto, we first had a speed of sixty inches per minute. That was too fast. Then we tried eighteen inches per minute. That was too slow. Finally, we settled on forty-four inches per minute. The idea is that a man must not be hurried in his work. He must have every second necessary but not a single unnecessary second. We have worked out speeds for each assembly, for the success of the chassis assembly caused us gradually to overhaul our entire method of manufacturing and to put all assembling in mechanically driven lines. The chassis assembling line, for instance, goes at a pace of six feet per minute; the front axle assembly line goes at one hundred eighty-nine inches per minute. In the chassis assembling are forty-five separate operations or stations. The first men fasten four mud-guard brackets to the chassis frame. Then the motor arrives on the tenth operation and so on in detail. Some men do only one or two small operations whereas others do more. The man who places a part does not fasten it because the part may not be fully in place until after several operations later. The man who puts in a bolt does not put on the nut. The man who puts on the nut does not tighten it. On operation number thirty-four, the budding motor gets its gasoline. It has previously received lubrication. On operation number forty-four, the radiator is filled with water, and on operation number forty-five the car drives out onto John R. Street.

6. **What is meant by the statement that they "had adopted the policy of 'man-high' work"?**

7. **Match the vocabulary term from the article to its description.**

	A device for raising or lowering some-thing	Electric generator	The framework on the car
fly-wheel magneto	○	○	○
chassis	○	○	○
windlass	○	○	○

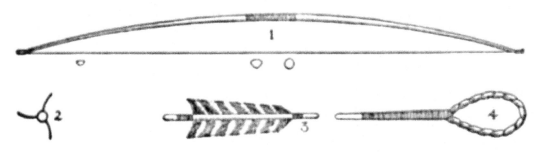

How to Make a Bow and Arrow

In early days, the Indian did not have the modern hunting rifle and was compelled to use bow and arrow in the chase. It is well for the modern boy to understand this weapon, for it can be made with ease and much good fun can be had with it. The Indian bow was short because though less efficient it was easier to carry than a long one, yet it did not lack power. We hear that many times Indians shot so hard that their arrows appeared on the far side of the animal, but the long bow, such was used by the old English archers or bowmen, was much the more powerful.

To make the bow, take a perfectly sound, straight, well-seasoned stick of about your own height and mark off a space as wide as your hand in the middle for a handle. This space should be left round and about an inch thick. The balance of the stick should be shaved down flat on one side for the front and rounded on the other for the back until it is about one inch wide and three-fourths of an inch thick near the handle tapering to about half that at the ends which are then notched for the cord. Next put on the cord and bend it until it is about five inches from the bow at the center. If one end bends more than the other, shave the other end until it becomes even.

After it is trimmed down to your strength, finish it with sand-paper. The best woods to use are apple, black walnut, slippery elm, mountain ash, or hickory.

More difficult to make than the bow is the arrow. The Indians made arrows of reeds and of straight shoots of arrow-wood or of elder, but we make better arrows out of hickory or ash. The arrow should be twenty-five inches long, round, and three-eighths of an inch thick. It should be notched at one end to take the bow-string, and just below this notch, it should have three feathers set around at equal intervals. These feathers are best made from either turkey or goose wings. With a sharp knife, cut a strip of the midrib on which is the vein; make three pieces, each two to three inches long. The Indians used to leave the midrib projecting at each end, and by these lash the feathers to the arrow without gluing, but it is easier to glue them and the arrows fly better. Indian arrow-heads were made of sharp flints or pieces of stone lashed to the arrowheads, but the best way to make them now is like the ferrule of an umbrella as this keeps the shaft from splitting. After this the arrow should be painted, both to keep it from warping and to make it easier to find in the forest by its bright color.

After the bow and arrows are made, one must learn to shoot with them. It is good to begin with the target close at hand, and gradually increase the distance as the archer becomes more expert. The Indians generally used their bows at short range so that it was easy to hit the mark, and considered rapid firing more important. In their competitions, the prize was given to him who should have the most arrows in the air at once, and it has been said that their record was eight.

8. **What is the meaning of a midrib as it is used in the sentence below?**

With a sharp, knife cut a strip of the midrib on which is the vein.

Ⓐ the elastic on the bow
Ⓑ the notch in the bow
Ⓒ the hairs of the feather
Ⓓ the stem of the feather

9. **The ferrule is found on the** _____ **of an object.**

How It Works

By attaching a small wheel to the end of a Morse-sounder lever, by arranging an ink-well for the wheel to dip into when the end falls, and by moving a paper ribbon slowly along for the wheel to press against when it rises, a self-recording Morse inker is produced. The ribbon-feeding apparatus is set in motion automatically by the current and continues to pull the ribbon along until the message is completed.

The Hughes type-printer covers a sheet of paper with printed characters in bold Roman type. The transmitter has a keyboard on which are marked letters, signs, and numbers. Also, a type-wheel, with the characters on its circumference, rotates by electricity. The receiver contains mechanisms for rotating another type-wheel synchronously—that is, in time—with the first, for shifting the wheel across the paper, for pressing the paper against the wheel, and for moving the paper when a fresh line is needed. These are too complicated to be described here in detail. By means of relays, one transmitter may be made to work five hundred receivers. In London a single operator controlling a keyboard in the central dispatching office causes typewritten messages to spell themselves out simultaneously in machines distributed all over the metropolis.

The tape machine resembles what was just described in many details. The main difference is that it prints on a continuous ribbon instead of on sheets.

Automatic electric printers of some kind or others are to be found in the foyers of all the principal hotels and clubs of our large cities and in the offices of bankers, stockbrokers, and newspaper editors. In London alone, over 500 million words are printed by the receivers in a year.

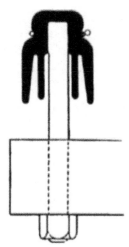

Fig. - Section of a telegraph wire insulator on its arm. The shaded circle is the line wire, the two blank circles indicate the wire which ties the line wire to the insulator.

10. What does the word "metropolis" mean in the sentence below?

In London a single operator controlling a keyboard in the central dispatching office causes type-written messages to spell themselves out simultaneously in machines distributed all over the metropolis.

Ⓐ Office
Ⓑ City
Ⓒ Telegraph
Ⓓ Desk

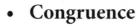

Congruence (Contd.)

1. **Which of the following angle relationships are congruent? Mark all that apply.**

   Ⓐ Vertical Angles
   Ⓑ Alternate Interior Angles
   Ⓒ Supplementary
   Ⓓ Complementary

2. **Find the measure of X.**

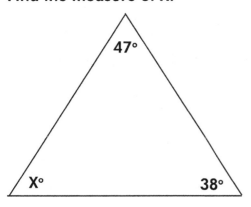

   Ⓐ 38°
   Ⓑ 47°
   Ⓒ 85°
   Ⓓ 95°

3. **If a triangle is an isosceles triangle, what is always true?**

   Ⓐ All of the angles are congruent
   Ⓑ The angles corresponding to the congruent sides are congruent
   Ⓒ The interior angles have a sum of 90°
   Ⓓ There are no angles or sides that are congruent

4. **Triangle ABC is an isosceles triangle with angle measure of 75° and 30°. What is the measure of the third angle?**

   Ⓐ 30°
   Ⓑ 45°
   Ⓒ 75°
   Ⓓ 105°

5. **Which of the following is not a way to define a triangle by the angle measures:**

Ⓐ Equiangular
Ⓑ Scalene
Ⓒ Acute
Ⓓ Right

Day 5 — Determine the Meaning of Words and Phrases (Contd.)

Patrick Henry Speech (March 23, 1775)

They tell us, sir, that we are weak; unable to cope with so formidable an adversary. But when shall we be stronger? Will it be the next week or the next year? Will it be when we are totally disarmed, and when a British guard shall be stationed in every house? Shall we gather strength by irresolution and inaction? Shall we acquire the means of effectual resistance by lying supinely on our backs and hugging the delusive phantom of hope, until our enemies shall have bound us hand and foot? Sir, we are not weak if we make a proper use of those means which the God of nature hath placed in our power. The millions of people, armed in the holy cause of liberty, and in such a country as that which we possess, are invincible by any force which our enemy can send against us. Besides, sir, we shall not fight our battles alone. There is a just God who presides over the destinies of nations, and who will raise up friends to fight our battles for us. The battle, sir, is not to the strong alone; it is to the vigilant, the active, the brave. Besides, sir, we have no election. If we were base enough to desire it, it is now too late to retire from the contest. There is no retreat but in submission and slavery! Our chains are forged! Their clanking may be heard on the plains of Boston! The war is inevitable—and let it come! I repeat it, sir, let it come.

It is in vain, sir, to extenuate the matter. Gentlemen may cry, Peace, Peace—but there is no peace. The war is actually begun! The next gale that sweeps from the north will bring to our ears the clash of resounding arms! Our brethren are already in the field! Why stand we here idle? What is it that gentlemen wish? What would they have? Is life so dear, or peace so sweet, as to be purchased at the price of chains and slavery? Forbid it, Almighty God! I know not what course others may take; but as for me, give me liberty or give me death!

6. **In the opening line of Paragraph 1, Henry uses the word "formidable" to describe his enemy, Great Britain. Based on the context of the word, what can you assume that "formidable" means?**

Ⓐ Impressively large and powerful
Ⓑ Mean
Ⓒ Aggressive
Ⓓ Not worthy of one's time

7. Henry uses _____ in the lines "Shall we acquire the means of effectual resistance by lying supinely on our backs and hugging the delusive phantom of hope" because it's clear he doesn't mean they should lie down on their backs and hang onto hope.

Ⓐ sarcasm
Ⓑ direct speech
Ⓒ a simile
Ⓓ a declarative statement

8. The series of questions that he asks the audience, most ardently reflect his desire for the colonial people to be:

Ⓐ Weak and docile
Ⓑ Angry and revolutionary
Ⓒ Strong but passive
Ⓓ Brave and stoic

9. The quote, "Is life so dear, or peace so sweet, as to be purchased at the price of chains and slavery?" uses imagery to compare freedom to remaining under British rule. Which type of imagery does he use?

Ⓐ Simile
Ⓑ Metaphor
Ⓒ Onomatopoeia
Ⓓ Assonance

10. What is the effect of his closing line changing from a series of questions to emphatic exclamation marks: "Forbid it, Almighty God! I know not what course others may take; but as for me, give me liberty or give me death!"

Ⓐ He stops questioning the people and emphatically demands that his answer is death over the slavery to the British people
Ⓑ The exclamations highlight how angry he is getting by the end of the speech.
Ⓒ Both of the above
Ⓓ None of the above

Challenge Yourself!

- Congruence

- Determine the Meaning of Words and Phrases

https://www.lumoslearning.com/a/dc9-15

Day 5

See Page 7 for Signup details

Self Defense Basics For Teens

While it's not nice to think about, the reality is that as we grow up and gain a bit more freedom, there are risks that come with being out in public. There can be weird people lurking out there when you catch public transport downtown, or even just hang out at your local mall. While hopefully, you'll never need it, a basic understanding of self-defense is a great way to keep yourself safe (and stop your parents from worrying!) Here are some self-defense tips for if you find yourself in an uncomfortable situation:

Pay attention

It seems simple, but you will automatically be less of a target if you are aware of your surroundings. That includes not using noise-canceling headphones or staring at your phone. It's a good way to avoid potentially dangerous situations, too, such as being alone in a train carriage at night.

Take advantage of your phone

While your phone can be your biggest distraction, it can also be your ultimate sidekick. Allow a few trusted friends and your parents access to your location (on an iPhone the app is Find Friends), and make sure to stay in contact via text whenever you're out.

Carry a whistle

If you ever are in a situation where you need to scare someone off or get a passerby's attention, a whistle is a great tool to have on hand. It will create enough of a commotion that you will hopefully be able to get out of there. Hopefully, you'll never need it, but it's better to be safe than sorry!

Brush up on some self-defense moves

It's never a bad idea to take a class to learn some moves. Karate, judo, boxing, or jiu-jitsu are all great sports that will give you a bit of experience and confidence in hand-to-hand combat. And who knows, maybe you'll end up finding your new favorite hobby!

Always remember that if you ever need to (and hopefully you don't), you are allowed to fight back against a potential attacker. Don't worry about being polite or making a scene, it is better to kick and scream and scratch to hopefully scare them away. The only time to avoid this is if the stranger is trying to fight you, then it's better to not engage and get out of there before it gets heated.

Week 3 - PSAT/NMSQT Prep

- Math
- Evidence Based Reading

https://www.lumoslearning.com/a/slh9-10

See Page 7 for Signup details

Weekly Fun Summer Photo Contest

Take a picture of your summer fun activity and share it on Twitter or Instagram

Use the **#SummerLearning** mention

@LumosLearning on Twitter or

@lumos.learning on Instagram

Tag friends and increase your chances of winning the contest

Participate and stand a chance to WIN $50 Amazon gift card!

Week 4 Summer Practice

Congruence (Contd.)

Day 1

1. **Which of the following is not a parallelogram?**

   Ⓐ Rectangle
   Ⓑ Rhombus
   Ⓒ Trapezoid
   Ⓓ Square

2. **Given ABCD is a parallelogram, find angle B.**

   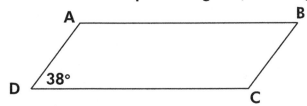

   Ⓐ 38°
   Ⓑ 52°
   Ⓒ 76°
   Ⓓ 142°

3. **What does it mean to be a regular polygon?**

   Ⓐ Each side is unique
   Ⓑ Each side is congruent to the others
   Ⓒ There are only seven sides to the shape
   Ⓓ It is closed

4. **Two students are discussing inscribed polygons. Eric says that if a triangle is inscribed in a circle that means the triangle is inside of the circle. John disagrees and says that if a triangle is inscribed in a circle that means the circle in inside of the triangle. Who is correct?**

5. What is the translation rule and the scale factor of the dilation as Circle F → Circle F´?

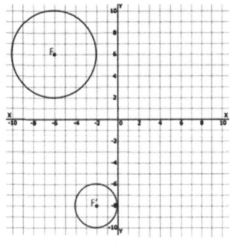

Ⓐ $(x,y) \rightarrow 2(x+4, y-14)$

Ⓑ $(x,y) \rightarrow \frac{1}{2}(x+4, y-14)$

Ⓒ $(x,y) \rightarrow \frac{1}{2}(x-4, y+14)$

Ⓓ $(x,y) \rightarrow \frac{1}{2}(x+14, y-4)$

Analyze in Detail an Author's Ideas

Day 1

Common Sense

Some writers have so been annoyed by society with the government as to leave little or no distinction between them whereas they are not only different but have different origins. Society is produced by our wants and the government by our wickedness. The former promotes our happiness positively by uniting our affections and the latter negatively by restraining our vices. The one encourages communication and the other creates differences. The first is a supporter, and the last is a punisher.

Society in every state is a blessing, but government even at its best state is but a necessary evil. In its worst state, it is an intolerable one. For when we suffer or are exposed to the same miseries by a government, our misfortune is heightened by reflecting that we furnish the means by which we suffer. Government is the badge of lost innocence, and the palaces of kings are built on the ruins of the places of paradise. If the impulses of a clear conscience, clear expectations, and irresistibly obeyed officials are there, then man would need no other lawgiver. However, that is not the case because he finds it necessary to surrender up a part of his property to furnish means for the protection of the rest. This, he is induced to do by the same cautiousness which in every other case advises him out of two evils to choose the least. Wherefore, security should be the true policy from the government, so that it follows whatever form appears most likely to ensure it to us with the least expense and greatest benefit is preferable to all others.

In order to gain a clear and just idea of the design of the government, let us suppose a small number of people settled in some sequestered part of the earth, unconnected with the rest, they will then represent the first peopling of any country or of the world. In this state of natural liberty, society will be their first thought. A thousand reasons will excite them thereto, the strength of one man is so unequal to his wants, and his mind so unfitted for continuous isolation, that he is soon obliged to seek assistance and relief of another. He requires the same. Four or five united people would be able to raise a tolerable dwelling in the midst of a wilderness, but one man might work out of the common period of life without accomplishing anything. When he had chopped for his timber, he could not remove it, nor build with it after it was removed. Therefore, hunger in the meantime would bother him from his work, and every different need calls him in a different way. Disease would disable him from living and reduce him to a state in which he might rather be said to perish than to die.

Thus, necessity would soon form our newly arrived emigrants into society the common blessings of the obligations of law and government unnecessary while they remained perfectly fair to each other. They prevail the first difficulties of emigration which bound them together in a common cause. They will begin to relax in their duty and attachment to each other. This is careless of establishing some form of government to supply the defect of moral virtue.

Some convenient tree will afford them a State-House, under the branches of which, the whole colony may assemble to deliberate on public matters. It is more than probable that their first laws will have the title only of regulations, and be enforced by no other penalty than public disfavor. In this first parliament every man, by natural right, will have a seat.

6. **What claim about the government is expressed by the author in the passage?**

   Ⓐ The government should look to other governments for structure.
   Ⓑ The government harms its citizens who work hard.
   Ⓒ The government lacks the proper protocol to function.
   Ⓓ The government spends inappropriately using its funds.

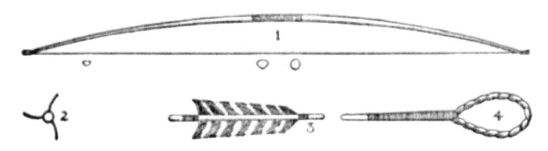

How to Make a Bow and Arrow

In early days, the Indian did not have the modern hunting rifle and was compelled to use bow and arrow in the chase. It is well for the modern boy to understand this weapon, for it can be made with ease and much good fun can be had with it. The Indian bow was short because though less efficient it was easier to carry than a long one, yet it did not lack power. We hear that many times Indians shot so hard that their arrows appeared on the far side of the animal, but the long bow, such was used by the old English archers or bowmen, was much the more powerful.

To make the bow, take a perfectly sound, straight, well-seasoned stick of about your own height and mark off a space as wide as your hand in the middle for a handle. This space should be left round and about an inch thick. The balance of the stick should be shaved down flat on one side for the front and rounded on the other for the back until it is about one inch wide and three-fourths of an inch thick near the handle tapering to about half that at the ends which are then notched for the cord. Next put on the cord and bend it until it is about five inches from the bow at the center. If one end bends more than the other, shave the other end until it becomes even.

After it is trimmed down to your strength, finish it with sand-paper. The best woods to use are apple, black walnut, slippery elm, mountain ash, or hickory.

More difficult to make than the bow is the arrow. The Indians made arrows of reeds and of straight shoots of arrow-wood or of elder, but we make better arrows out of hickory or ash. The arrow should be twenty-five inches long, round, and three-eighths of an inch thick. It should be notched at one end to take the bow-string, and just below this notch, it should have three feathers set around at equal intervals. These feathers are best made from either turkey or goose wings. With a sharp knife, cut a strip of the midrib on which is the vein; make three pieces, each two to three inches long. The Indians used to leave the midrib projecting at each end, and by these lash the feathers to the arrow without gluing, but it is easier to glue them and the arrows fly better. Indian arrow-heads were made of sharp flints or pieces of stone lashed to the arrowheads, but the best way to make them now is like the ferrule of an umbrella as this keeps the shaft from splitting. After this the arrow should be painted, both to keep it from warping and to make it easier to find in the forest by its bright color.

After the bow and arrows are made, one must learn to shoot with them. It is good to begin with the target close at hand, and gradually increase the distance as the archer becomes more expert. The Indians generally used their bows at short range so that it was easy to hit the mark, and considered rapid firing more important. In their competitions, the prize was given to him who should have the most arrows in the air at once, and it has been said that their record was eight.

7. **In what way are the author's ideas developed?**

   Ⓐ By explaining the materials that the Indians used.
   Ⓑ By systematically detailing the steps of making and shooting an arrow from a bow.
   Ⓒ By describing the difference between making the bow and making the arrow.
   Ⓓ By detailing the importance of why the Indian used such a powerful bow and arrow.

How It Works

By attaching a small wheel to the end of a Morse-sounder lever, by arranging an ink-well for the wheel to dip into when the end falls, and by moving a paper ribbon slowly along for the wheel to press against when it rises, a self-recording Morse inker is produced. The ribbon-feeding apparatus is set in motion automatically by the current and continues to pull the ribbon along until the message is completed.

The Hughes type-printer covers a sheet of paper with printed characters in bold Roman type. The transmitter has a keyboard on which are marked letters, signs, and numbers. Also, a type-wheel, with the characters on its circumference, rotates by electricity. The receiver contains mechanisms for rotating another type-wheel synchronously—that is, in time—with the first, for shifting the wheel across the paper, for pressing the paper against the wheel, and for moving the paper when a fresh line is needed. These are too complicated to be described here in detail. By means of relays, one transmitter may be made to work five hundred receivers. In London a single operator controlling a keyboard in the central dispatching office causes typewritten messages to spell themselves out simultaneously in machines distributed all over the metropolis.

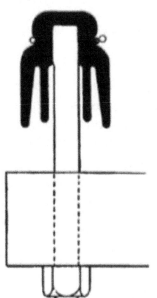

The tape machine resembles what was just described in many details. The main difference is that it prints on a continuous ribbon instead of on sheets.

Automatic electric printers of some kind or others are to be found in the foyers of all the principal hotels and clubs of our large cities and in the offices of bankers, stockbrokers, and newspaper editors. In London alone, over 500 million words are printed by the receivers in a year.

Fig. - Section of a telegraph wire insulator on its arm. The shaded circle is the line wire, the two blank circles indicate the wire which ties the line wire to the insulator.

8. **How is the author's claim about the importance of the wireless telegraph developed?**

Whales, Dolphins, and Porpoises of the Western North Atlantic

All five species of large whales with a dorsal fin belong to the same major baleen whale group, the balaenopterid whales or rorquals. All are characterized by the presence of a series of ventral grooves usually visible on stranded specimens and the length and number of which are diagnostic to species. In addition, all species, with the exception of the humpback whale, have at least one distinctive (though often not prominent) ridge along the head from just in front of the blowhole to near the tip of the snout. (The humpback whale, on the other hand, is distinguished by numerous knobs, some of which are located along the line of the head ridge with others scattered on the top of the head.) In Bryde's whale, the single head ridge characteristic of the other rorquals is supplemented by two auxiliary ridges one on each side of the main ridge.

At sea, these whales often appear very similar and must be examined carefully before they can be reliably identified.

In general, though the characteristics of behavior may vary from one encounter to the next, based on the activities in which the animal is engaged. Whales in this group may be distinguished from each other on the basis of differences in 1) the size, shape, and position of the dorsal fin and the timing of its appearance on the surface relative to the animal's blow (in general, the larger the whale, the smaller the dorsal fin—the further back its position and the later its appearance on the surface after the animal's blow); 2) the height of the body in the area of the dorsal fin, relative to the size of the dorsal fin, which is exposed as the animal sounds; 3) sometimes the blow rate and movement patterns; and 4) the shape and color of the head.

Despite variability in behavior by members of the same species from one encounter to the next, an observer can greatly increase the reliability of his identification by forming the habit of working systematically through a set of characteristics for the species rather than depending on any single characteristic.

There are three species of large whales without a dorsal fin in the western North Atlantic Ocean. Two of these, the bowhead or Greenland whale, and its more widely distributed close relative, the right whale, are baleen whales. The third, the sperm whale, is a toothed whale. The first two have relatively smooth backs without even a trace of a dorsal fin. The sperm whale has a humplike low, thick, dorsal ridge, which from certain views particularly when the animal is humping up to begin a dive may be clearly visible and looks like a fin. But because the profile of that hump and the knuckles which follow it are often not very prominent in this species, it has been classified with the finless big whales.

All three species are characterized by very distinctive blows or spouts. In both the bowhead and the right whales, the projection of the blow upward from two widely separated blowholes assumes a very wide V-shape with two distinct columns, which may be seen when the animals are viewed from front or back. Though this character may be visible under ideal conditions in many of the other baleen whales species as well, it is exaggerated and uniformly distinct in the bowhead and right whales and may be used as one of the primary key characters. In the sperm whale, the blow emanates from a blowhole which is displaced to the left of the head near the front and projects obliquely forward to the animal's left. This blow seen under ideal conditions positively labels a large whale as a sperm whale.

9. What claim does the author indirectly make about different whales?

The True Story of Our National Calamity of Flood, Fire and Tornado

With the rest of the country, fair Dayton sorrowed for Omaha. Two days later Omaha, bowed and almost broken by her own misfortune, looked with sympathy across to Dayton, whose woe was even greater. A thousand communities in the United States read the story and in their own sense of security sent eager volunteers of assistance to the stricken districts. Not one of them has the assurance that it may not be next. There is no sure definition of the course of the earthquake, the path of the wind, or the time and place of the storm-cloud. Science has its limitations.

In the legal jargon of the practice of torts, such occurrences as these are known as "acts of God." Theologians who attempt to solve the mysteries of Providence have found in such occasions the evidence of Divine wrath and warning to the smitten people. But to seek the reason and to know the purpose, if there be purpose in it, is not necessary. The fact is enough. It challenges, staggers, calls a halt, compels men and women to think—and even to pray.

But the flood did not confine itself to Dayton. It laid its watery hand of death and destruction over a whole tier of states from the Great Lakes to New England, and over the vast area to the southward which is veined by the Ohio River and its tributaries, and extending from the Mississippi Valley almost to the Atlantic seaboard. And as this awful deluge drained from the land into Nature's watercourses the demons of death and devastation danced attendance on its mad rush that laid waste the border-lands of the Mississippi River from Illinois to the Gulf of Mexico.

Those who have never seen a great flood do not know the meaning of the Scriptural phrase, "the abomination of desolation."

An explosion, a railroad wreck, even a fire—these are bad enough in their pictorial effect of shattered ruins and confusion. But for giving one an oppressive sense of death-like misery, there is nothing equal to a flood.

I do not speak now of the loss of life, which is unspeakably dreadful but of the scenic effect of the disaster. It just grips and benumbs you with its awfulness.

In the flat country of the Middle West, there is less likelihood of swift, complete destruction than in narrow valleys, like those of Johnstown and Austin in Pennsylvania. But the effect is, if anything, more gruesome.

After the crest has passed there are miles and miles of inundated land, with only trees and half- submerged buildings and floating wreckage to break the monotony; just a vast lake of yellow, muddy water, swirling and boiling as it seeks to find its level.

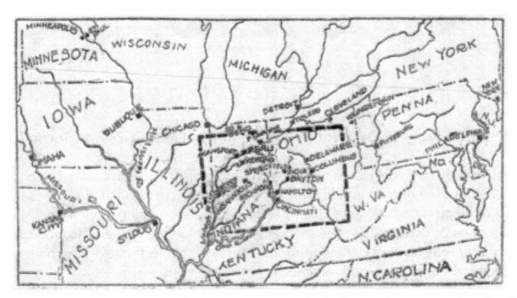

The cities and towns enclosed by the heavy black dotted lines were the chief sufferers by the sweep of waters.

10. What does the author do to develop his ideas about the destruction of flooding?

Challenge Yourself!

- **Congruence**
- **Analyze in Detail an Author's Ideas**

https://www.lumoslearning.com/a/dc9-16

Day 1

See Page 7 for Signup details

Circles

1. Quadrilateral ABCD is inscribed in circle O, as shown in the figure below. What is m∠DAB?

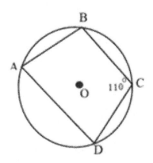

   Ⓐ 60°
   Ⓑ 70°
   Ⓒ 80°
   Ⓓ 82°

2. What is m$\widehat{VW}$ in the figure below?

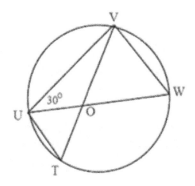

   Ⓐ 60°
   Ⓑ 30°
   Ⓒ 45°
   Ⓓ 90°

3. A circle is inscribed inside ΔXYZ with tangent points U, V, and W. What is the length of segment XY?

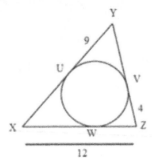

Ⓐ 60°
Ⓑ 70°
Ⓒ 80°
Ⓓ 82°

4. A circle is inscribed inside ΔUVW with tangent points X,Y,and Z. What is the length of segment UW?

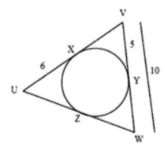

Ⓐ 11
Ⓑ 12
Ⓒ 5
Ⓓ 10

5. The figure below shows circle O, with segment QR tangent to circle O at point Q. The figure also shows that x is the length of the radius of circle O. What is the value of x?

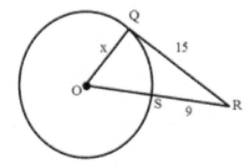

Ⓐ 13
Ⓑ 15
Ⓒ 8
Ⓓ 7

Determine an Author's Point of View

Common Sense

Some writers have so been annoyed by society with the government as to leave little or no distinction between them whereas they are not only different but have different origins. Society is produced by our wants and the government by our wickedness. The former promotes our happiness positively by uniting our affections and the latter negatively by restraining our vices. The one encourages communication and the other creates differences. The first is a supporter, and the last is a punisher.

Society in every state is a blessing, but government even at its best state is but a necessary evil. In its worst state, it is an intolerable one. For when we suffer or are exposed to the same miseries by a government, our misfortune is heightened by reflecting that we furnish the means by which we suffer. Government is the badge of lost innocence, and the palaces of kings are built on the ruins of the places of paradise. If the impulses of a clear conscience, clear expectations, and irresistibly obeyed officials are there, then man would need no other lawgiver. However, that is not the case because he finds it necessary to surrender up a part of his property to furnish means for the protection of the rest. This, he is induced to do by the same cautiousness which in every other case advises him out of two evils to choose the least. Wherefore, security should be the true policy from the government, so that it follows whatever form appears most likely to ensure it to us with the least expense and greatest benefit is preferable to all others.

In order to gain a clear and just idea of the design of the government, let us suppose a small number of people settled in some sequestered part of the earth, unconnected with the rest, they will then represent the first peopling of any country or of the world. In this state of natural liberty, society will be their first thought. A thousand reasons will excite them thereto, the strength of one man is so unequal to his wants, and his mind so unfitted for continuous isolation, that he is soon obliged to seek assistance and relief of another. He requires the same. Four or five united people would be able to raise a tolerable dwelling in the midst of a wilderness, but one man might work out of the common period of life without accomplishing anything. When he had chopped for his timber, he could not remove it, nor build with it after it was removed. Therefore, hunger in the meantime would bother him from his work, and every different need calls him in a different way. Disease would disable him from living and reduce him to a state in which he might rather be said to perish than to die.

Thus, necessity would soon form our newly arrived emigrants into society the common blessings of the obligations of law and government unnecessary while they remained perfectly fair to each other. They prevail the first difficulties of emigration which bound them together in a common cause. They will begin to relax in their duty and attachment to each other. This is careless of establishing some form of government to supply the defect of moral virtue.

Some convenient tree will afford them a State-House, under the branches of which, the whole colony may assemble to deliberate on public matters. It is more than probable that their first laws will have the title only of regulations, and be enforced by no other penalty than public disfavor. In this first parliament every man, by natural right, will have a seat.

6. What were the results of Thomas Paine writing "Common Sense"?

My Life and Work

I believe that this was the first moving line ever installed. The idea came in a general way from the overhead trolley that the Chicago packers use in packaging beef. We had previously assembled the fly-wheel magneto in the usual method. With one workman doing a complete job, he could turn out from thirty-five to forty pieces in a nine-hour day or about twenty minutes to an assembly. What he did alone was then spread into twenty-nine operations that cut down the assembly time to thirteen minutes, ten seconds. Then we raised the height of the line eight inches which was in 1914 and cut the time to seven minutes. Further experimenting with the speed that the work should move quickly and cut the time down to five minutes. In short, the result is the aid of scientific study one man is now able to do somewhat more than four did only comparatively a few years ago. That line established the efficiency of the method, and we now use it everywhere. The assembling of the motor, formerly done by one man, is now divided into eighty-four operations. Those men do the work that three times their number formerly did. In a short time, we tried out the plan on the framework on the car or the chassis.

About the best we had done in stationary chassis assembling was an average of twelve hours and twenty-eight minutes per chassis. We tried the experiment of drawing the chassis with a rope and windlass down a line two hundred fifty feet long. Six assemblers traveled with the chassis and picked up the parts from piles placed along the line. This rough experiment reduced the time to five hours fifty minutes per chassis. In the early part of 1914, we elevated the assembly line. We had adopted the policy of "man-high" work. We had one line twenty-six and three-quarter inches and another twenty-four and one-half inches from the floor to suit squads of different heights. The waist-high arrangement and a further subdivision of work so that each man had fewer movements cut down the labor time per chassis to one hour thirty-three minutes. Only the chassis was then assembled in the line. The body was placed on "John R. Street" which is the famous street that runs through our Highland Park factories. Now the line assembles the whole car.

It must not be imagined, however, that all this worked out as quickly as it sounds. The speed of the moving work had to be carefully tried out. In the fly-wheel magneto, we first had a speed of sixty inches per minute. That was too fast. Then we tried eighteen inches per minute. That was too slow. Finally, we settled on forty-four inches per minute. The idea is that a man must not be hurried in his work. He must have every second necessary but not a single unnecessary second. We have worked out speeds for each assembly, for the success of the chassis assembly caused us gradually to overhaul our entire method of manufacturing and to put all assembling in mechanically driven lines. The chassis assembling line, for instance, goes at a pace of six feet per minute; the front axle assembly line goes at one hundred eighty-nine inches per minute. In the chassis assembling are forty-five separate operations or stations. The first men fasten four mud-guard brackets to the chassis frame. Then the motor arrives on the tenth operation and so on in detail. Some men do only one or two small operations whereas others do more. The man who places a part does not fasten it because the part may not be fully in place until after several operations later. The man who puts in a bolt does not put on the nut. The man who puts on the nut does not tighten it. On operation number thirty-four, the budding motor gets its gasoline. It has previously received lubrication. On operation number forty-four, the radiator is filled with water, and on operation number forty-five the car drives out onto John R. Street.

7. **What is the author's purpose for showing how long each part of the assembly line takes?**

Ⓐ to allow the reader to visualize each step in the assembly line
Ⓑ for the method to be replicated
Ⓒ to show in detail how his assembly line worked
Ⓓ to provide details for assembling a car

8. **Why does the author describe how the men on the line do not complete the process before sending that part of the car down the assembly line?**

Declaration of Independence

The history of the present King of Great Britain is a history of repeated injuries and seizures, all having in direct object the establishment of an absolute Tyranny over these States. To prove this, let Facts be submitted to a candid world.

He has refused his Assent to Laws, the most wholesome and necessary for the public good.

He has forbidden his Governors to pass Laws of immediate and pressing importance unless suspended in their operation till his Assent should be obtained; and when so suspended, he has utterly neglected to attend to them.

He has refused to pass other Laws for the accommodation of large districts of people, unless those people would relinquish the right of Representation in the Legislature, a right inestimable to them and formidable to tyrants only.

He has called together legislative bodies at places unusual, uncomfortable, and distant from the depository of their Public Records, for the sole purpose of fatiguing them into compliance with his measures.

He has dissolved Representative Houses repeatedly, for opposing with manly firmness his invasions on the rights of the people.

He has refused for a long time, after such dissolutions, to cause others to be elected; whereby the Legislative Powers, incapable of Annihilation, have returned to the People at large for their exercise; the State remaining in the meantime exposed to all the dangers of invasion from without, and convulsions within.

He has endeavored to prevent the population of these States; for that purpose obstructing the Laws of Naturalization of Foreigners; refusing to pass others to encourage their migration hither, and raising the conditions of new Appropriations of Lands.

9. What word best describes Jefferson's attitude and point of view towards the British government?

Ⓐ Bitter
Ⓑ Callous
Ⓒ Detached
Ⓓ Egotistical

How to Make a Bow and Arrow

In early days, the Indian did not have the modern hunting rifle and was compelled to use bow and arrow in the chase. It is well for the modern boy to understand this weapon, for it can be made with ease and much good fun can be had with it. The Indian bow was short because though less efficient it was easier to carry than a long one, yet it did not lack power. We hear that many times Indians shot so hard that their arrows appeared on the far side of the animal, but the long bow, such was used by the old English archers or bowmen, was much the more powerful.

To make the bow, take a perfectly sound, straight, well-seasoned stick of about your own height and mark off a space as wide as your hand in the middle for a handle. This space should be left round and about an inch thick. The balance of the stick should be shaved down flat on one side for the front and rounded on the other for the back until it is about one inch wide and three-fourths of an inch thick near the handle tapering to about half that at the ends which are then notched for the cord. Next put on the cord and bend it until it is about five inches from the bow at the center. If one end bends more than the other, shave the other end until it becomes even.

After it is trimmed down to your strength, finish it with sand-paper. The best woods to use are apple, black walnut, slippery elm, mountain ash, or hickory.

More difficult to make than the bow is the arrow. The Indians made arrows of reeds and of straight shoots of arrow-wood or of elder, but we make better arrows out of hickory or ash. The arrow should be twenty-five inches long, round, and three-eighths of an inch thick. It should be notched at one end to take the bow-string, and just below this notch, it should have three feathers set around at equal intervals. These feathers are best made from either turkey or goose wings. With a sharp knife, cut a strip of the midrib on which is the vein; make three pieces, each two to three inches long. The Indians used to leave the midrib projecting at each end, and by these lash the feathers to the arrow without gluing, but it is easier to glue them and the arrows fly better. Indian arrow-heads were made of sharp flints or pieces of stone lashed to the arrowheads, but the best way to make them now is like the ferrule of an umbrella as this keeps the shaft from splitting. After this the arrow should be painted, both to keep it from warping and to make it easier to find in the forest by its bright color.

After the bow and arrows are made, one must learn to shoot with them. It is good to begin with the target close at hand, and gradually increase the distance as the archer becomes more expert. The Indians generally used their bows at short range so that it was easy to hit the mark, and considered rapid firing more important. In their competitions, the prize was given to him who should have the most arrows in the air at once, and it has been said that their record was eight.

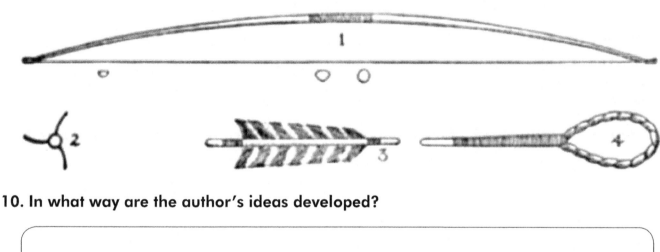

10. In what way are the author's ideas developed?

Challenge Yourself!

- Circles

- Determine an author's point of view

https://www.lumoslearning.com/a/dc9-17

Day 2

See Page 7 for Signup details

Day 3

1. What is the approximate area of the sector of the circle that is shaded? Use $\pi=3.14$ in your calculations.

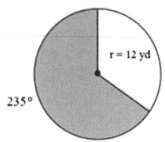

Ⓐ 295.16 yd
Ⓑ 94π yd²
Ⓒ 295.16 yd²
Ⓓ 24.597 yd²

2. What is the exact length of the arc subtended by the sector of the circle that is shaded?

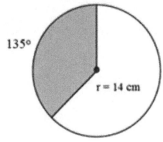

Ⓐ $\frac{21}{2}$ cm

Ⓑ $\frac{21}{2}\pi$ cm

Ⓒ $\frac{21}{4}\pi$ cm

Ⓓ $\frac{21}{2}\pi$ cm²

3. A circle has its center at the point $(-4,-7)$ and passes through the point $(4,-9)$. Which one of the following points lies on the circle?

Ⓐ $(-11,-4)$
Ⓑ $(-12,-5)$
Ⓒ $(5,-8)$
Ⓓ $(-5,-12)$

4. A circle has its center at the point (−3,8) and passes through the point (4,3). Which one of the following points lies on the circle?

Ⓐ (2,15)
Ⓑ (−10,12)
Ⓒ (3,14)
Ⓓ (−9,15)

5. What is the equation, in slope-intercept form, of the line that passes through the point (3,1), and is parallel to the line with the equation $y=\frac{2}{5}x-7$?

Ⓐ $y-1=\frac{2}{5}(x-3)$

Ⓑ $y-1=\frac{2}{5}x-\frac{6}{5}$

Ⓒ $y=\frac{2}{5}x-\frac{1}{5}$

Ⓓ $y-1=\frac{2}{5}x+1$

Day 3

Analyze Various Accounts of a Subject

Declaration of Independence

The history of the present King of Great Britain is a history of repeated injuries and seizures, all having in direct object the establishment of an absolute Tyranny over these States. To prove this, let Facts be submitted to a candid world.

He has refused his Assent to Laws, the most wholesome and necessary for the public good.

He has forbidden his Governors to pass Laws of immediate and pressing importance unless suspended in their operation till his Assent should be obtained; and when so suspended, he has utterly neglected to attend to them.

He has refused to pass other Laws for the accommodation of large districts of people, unless those people would relinquish the right of Representation in the Legislature, a right inestimable to them and formidable to tyrants only.

He has called together legislative bodies at places unusual, uncomfortable, and distant from the depository of their Public Records, for the sole purpose of fatiguing them into compliance with his measures.

He has dissolved Representative Houses repeatedly, for opposing with manly firmness his invasions on the rights of the people.

He has refused for a long time, after such dissolutions, to cause others to be elected; whereby the Legislative Powers, incapable of Annihilation, have returned to the People at large for their exercise; the State remaining in the meantime exposed to all the dangers of invasion from without, and convulsions within.

He has endeavored to prevent the population of these States; for that purpose obstructing the Laws of Naturalization of Foreigners; refusing to pass others to encourage their migration hither, and raising the conditions of new Appropriations of Lands.

6. **Which words in the rough draft seem to be the most important based on their appearance?**

   Ⓐ When in the course of human events
   Ⓑ United States of America
   Ⓒ Declaration
   Ⓓ Independence

7. **Comparing the rough draft with the published copy of the "Declaration of Independence," Jefferson made many _____.**

8. **Most likely the most difficult section to write using the rough draft would be _____ _____.**

How to Make a Bow and Arrow

In early days, the Indian did not have the modern hunting rifle and was compelled to use bow and arrow in the chase. It is well for the modern boy to understand this weapon, for it can be made with ease and much good fun can be had with it. The Indian bow was short because though less efficient it was easier to carry than a long one, yet it did not lack power. We hear that many times Indians shot so hard that their arrows appeared on the far side of the animal, but the long bow, such was used by the old English archers or bowmen, was much the more powerful.

To make the bow, take a perfectly sound, straight, well-seasoned stick of about your own height and mark off a space as wide as your hand in the middle for a handle. This space should be left round and about an inch thick. The balance of the stick should be shaved down flat on one side for the front and rounded on the other for the back until it is about one inch wide and three-fourths of an inch thick near the handle tapering to about half that at the ends which are then notched for the cord. Next put on the cord and bend it until it is about five inches from the bow at the center. If one end bends more than the other, shave the other end until it becomes even.

After it is trimmed down to your strength, finish it with sand-paper. The best woods to use are apple, black walnut, slippery elm, mountain ash, or hickory.

More difficult to make than the bow is the arrow. The Indians made arrows of reeds and of straight shoots of arrow-wood or of elder, but we make better arrows out of hickory or ash. The arrow should be twenty-five inches long, round, and three-eighths of an inch thick. It should be notched at one end to take the bow-string, and just below this notch, it should have three feathers set around at equal intervals. These feathers are best made from either turkey or goose wings. With a sharp knife, cut a strip of the midrib on which is the vein; make three pieces, each two to three inches long. The Indians used to leave the midrib projecting at each end, and by these lash the feathers to the arrow without gluing, but it is easier to glue them and the arrows fly better. Indian arrow-heads were made of sharp flints or pieces of stone lashed to the arrowheads, but the best way to make them now is like the ferrule of an umbrella as this keeps the shaft from splitting. After this the arrow should be painted, both to keep it from warping and to make it easier to find in the forest by its bright color.

After the bow and arrows are made, one must learn to shoot with them. It is good to begin with the target close at hand, and gradually increase the distance as the archer becomes more expert. The Indians generally used their bows at short range so that it was easy to hit the mark, and considered rapid firing more important. In their competitions, the prize was given to him who should have the most arrows in the air at once, and it has been said that their record was eight.

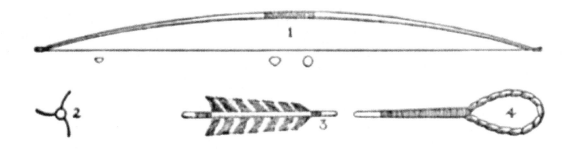

9. Why is there a shaded part in the drawing for the bow (1)?

***

How It Works

By attaching a small wheel to the end of a Morse-sounder lever, by arranging an ink-well for the wheel to dip into when the end falls, and by moving a paper ribbon slowly along for the wheel to press against when it rises, a self-recording Morse inker is produced. The ribbon-feeding apparatus is set in motion automatically by the current and continues to pull the ribbon along until the message is completed.

The Hughes type-printer covers a sheet of paper with printed characters in bold Roman type. The transmitter has a keyboard on which are marked letters, signs, and numbers. Also, a type-wheel, with the characters on its circumference, rotates by electricity. The receiver contains mechanisms for rotating another type-wheel synchronously—that is, in time—with the first, for shifting the wheel across the paper, for pressing the paper against the wheel, and for moving the paper when a fresh line is needed. These are too complicated to be described here in detail. By means of relays, one transmitter may be made to work five hundred receivers. In London a single operator controlling a keyboard in the central dispatching office causes typewritten messages to spell themselves out simultaneously in machines distributed all over the metropolis.

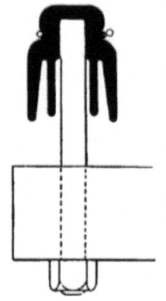

The tape machine resembles what was just described in many details. The main difference is that it prints on a continuous ribbon instead of on sheets.

Automatic electric printers of some kind or others are to be found in the foyers of all the principal hotels and clubs of our large cities and in the offices of bankers, stockbrokers, and newspaper editors. In London alone, over 500 million words are printed by the receivers in a year.

Fig. - Section of a telegraph wire insulator on its arm. The shaded circle is the line wire, the two blank circles indicate the wire which ties the line wire to the insulator.

10. Observe the drawing at the end of the passage.

What does the drawing reveal about the wireless telegraph?

1. What value on the number line in the figure below divides segment AB into two parts having a ratio of their lengths of 2:1?

Ⓐ 8
Ⓑ 9
Ⓒ 5
Ⓓ 11

2. What value on the number line in the figure below divides segment CD into two parts having a ratio of their lengths of 1:2?

Ⓐ -20
Ⓑ 10
Ⓒ -10
Ⓓ -5

3. What is the perimeter of polygon EFGH in the figure below?

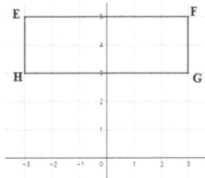

Ⓐ 12
Ⓑ 14
Ⓒ 16
Ⓓ 18

4. What is the volume of the prism shown below?

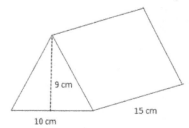

9 cm

10 cm

15 cm

- (A) 1350 cm³
- (B) 1350 cm
- (C) 675 cm³
- (D) 675cm

5. Andre wants to fill a cylindrical tank with water. If the tank has a diameter of 2 meters and is 3 meters high, how much water will Andre need? If necessary, use $\pi=3.14$

- (A) 8.14 m³
- (B) 9.14 m³
- (C) 9.42 m³
- (D) 10.25 m³

Day 4

Delineate and Evaluate the Argument

Can bullying be overcome by Kindness

Being Kind is not easy. In fact, it is very complex. If kindness had been a simple behavioral trait, then everyone would have been kind, and no one would have experienced meanness or bullying. A world in which Kindness is the norm is an ideal world. When we ask if it is possible to have homes, schools, communities where Kindness is the norm, the answer would be Yes. However to do so, we need to teach, model, and reward kindness.

For being kind, one needs to think about the needs and concerns of others. Inculcating the behavior of volunteering to help others and work for that affect their communities helps in developing Kindness and empathy. Compassionate thinking and generous actions demonstrate kindness.

Unfortunately, in many schools, negative behaviors such as bullying results in punishment which is thought to reduce this kind of behaviour in future. On the contrary, research shows that for "zero-tolerance" and to end bullying and violence punishment-based approaches do not work. Given this knowledge, it makes better sense to focus on teaching and modeling behaviors such as kindness and empathy.

Ways to Teach Kindness
• Mindfulness involves becoming aware of the specific thought, emotion, or behavior. This means that by being mentally flexible, and through training, even young children can learn kindness.

• Social-Emotional Learning (SEL) teaches kindness by focusing on cooperation, responsibility, self-control, empathy, and provides specific actions to build these skills.

• Acts of Kindness are actions such as doing something nice to others. Doing acts of kindness cause positive ripple effects to those who experience and witness kindness.

Impact of Teaching Kindness
Elementary school students who performed three acts of kindness per week saw that they were significantly more accepted by their peers compared to kids who did not perform three kind acts of kindness. Students who are taught kindness are more empathic, more socially aware, and connected. They also receive higher grades. Be kind—it is free, and the payback is good for all!

6. **Logical flaws in an argument are called logical...**

   Ⓐ Mistakes
   Ⓑ Fallacies
   Ⓒ Errors
   Ⓓ Buttons

7. **An example of a logical fallacy is**

   Ⓐ Begging the question
   Ⓑ Slippery slope
   Ⓒ Straw man
   Ⓓ All of the above

8. **This type of logical fallacy, can best be defined as: I didn't remember to take out the trash, so now it is going to be out there all week. My next door neighbors will get angry at the smell of it and probably start a campaign to kick me out of here.**

   Ⓐ Begging the question
   Ⓑ Slippery slope
   Ⓒ Straw man
   Ⓓ Ad hominem

9. **The flaw in this argument is: "Students who are taught kindness are more empathic, more socially aware and connected. They also receive higher grades."**

   Ⓐ Begging the question
   Ⓑ Slippery slope
   Ⓒ Straw man
   Ⓓ Ad hominem

10. In argument making, one often has to recognize the counter argument. An example of counterargument in the essay can be found in which line? Is it effective and why?

1. The cylinder shown below is sliced. Which of the following shapes could be a result?

Ⓐ a triangle
Ⓑ a circle
Ⓒ a square
Ⓓ both A and B

2. Which shape could result in an equilateral triangle if a cross section were taken?

Ⓐ

Ⓑ

Ⓒ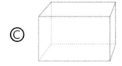

Ⓓ

3. What is the area of a square with a perimeter of 4 feet?

Ⓐ 1 ft
Ⓑ 1 ft²
Ⓒ 2 ft²
Ⓓ 4 ft²

4. John is painting a fence around a rectangular yard. The house forms one side of the fencing and will not be painted. The yard is 30 feet long and 20 feet wide. If the fence is 3 feet high and a gallon of paint covers 50 ft², how many gallons should John purchase.

  Ⓐ 3 gallons
  Ⓑ 4 gallons
  Ⓒ 5 gallons
  Ⓓ Cannot be found from the information given

5. Cynthia is attempting to find the volume of a tree trunk. She knows the circumference is 5π feet. If the trunk is about 10 feet tall, what is the best estimate for the volume of the trunk?

  Ⓐ 2.5 feet³
  Ⓑ 62.5π feet³
  Ⓒ 6.2π feet³
  Ⓓ 25 feet³

Textual Evidence to Support Analysis

Day 5

Lower the Voting Age

When we think about the minimum age for voting, the first question that comes to our mind is: How young is too young? The minimum ages vary in different context such as that for driving, marriage, joining the army, and so on. While there is an argument that 16 and 17-year-olds are too immature to vote, there is also a danger that they might not vote at all.

When we look at the trends in voting turnout, it is a bit distressing to note the numbers. This disenchantment of the young towards voting is a matter of concern as voting is a habit and it is necessary to ensure that this habit is inculcated as soon as possible. If left unattended, it could lead to ever lower participation rates in the decades to come, questioning the legitimacy of governments in a vicious spiral in which the poor voting turnout results in skepticism towards democracy, and vice versa.

The causes of disillusionment are many. Young adults look at voting as a privilege or choice rather than duty. The politically active persons tend to campaign on single issues and matters of concern rather than for a particular party. Politicians also focus and woo the older voters rather than the young as they are more likely to vote.

There are some countries which have made voting compulsory thereby increasing turnout rates. But this is not a solution to the disillusionment. Governments have to focus on ways to rekindle the passion, rather than continue to ignore the absence of it. A good step would be to lower the voting age to 16. This would ensure that new voters get off to the best possible start.

This cannot be considered as an arbitrary change. The current age for voting, which is 18 coincides with finishing compulsory education and leaving home. Away from their parents, they have no established voters to follow as they have limited connection to their new communities. In this process, they remain away from the electoral roll, and the habit of voting is not established. If the voting age is 16 years, they can get into the habit of voting by accompanying their parents to polling stations.

However, just lowering the voting age alone will not help. Schools can contribute to better turnouts by helping the children register. Governments also need to put more effort into keeping electoral rolls current. Civics lessons can be improved. Schools can also have courses that promote open debate thereby giving the pupils a chance to vote in aspects of their school. This will boost political commitment later in life.

A lower voting age would strengthen the voice of the young and also signal that their opinions matter. They, as future citizens would be the ones who would be facing issues such as climate change, pensions, and healthcare, etc. in the future.

6. **Implicit reading forces the reader to**

Ⓐ Read between the lines and determine the purpose of the writer's work.
Ⓑ Use context clues to understand what words mean.
Ⓒ Make inferences based on the reasoning present.
Ⓓ All of the above.

7. **In the nonfiction article, "Lowering the Voting Age" the author uses which of the following points to support her argument?**

Ⓐ After a child turns 18, they are usually no longer with their parents in the home, and won't have the right modeling to vote.
Ⓑ Schools should be helping to enroll children in voter registration and at 18, these students are no longer in schools.
Ⓒ Lowering the voting age would give young people a voice in their communities.
Ⓓ All of the above.

8. **In which paragraph(s) does the author provide evidence that supports the effectiveness of lowering the voting age?**

Ⓐ In paragraph 2
Ⓑ In paragraph 3
Ⓒ In the conclusion
Ⓓ She offers none

9. **How can the writer improve their conclusion paragraph?**

Ⓐ It's perfect the way it is.
Ⓑ She should recap the main points of her argument in the conclusion.
Ⓒ She should use better punctuation.
Ⓓ She needs to write at least 8 more sentences.

10. **Which sentence in paragraph 1 best illustrates the author's thesis:**

Challenge Yourself!

- **Geometry**
- **Textual Evidence to Support Analysis**

https://www.lumoslearning.com/a/dc9-20

Day 5

See Page 7 for Signup details

Learn To Code In Your Summer Break

So much of our life is online now, and you probably spend a lot of time looking at coded websites and apps... so if you can't beat 'em, join 'em! Learning to code is a great skill to immerse you in tech and get you thinking. You could design your own app or blog, and along the way will learn problem-solving and critical thinking skills that will help you in school and beyond.

Coding is so popular, in fact, that it can be hard to know where to start your learning! Here are some of our favorite coding resources for teens, try a few and see which one works best for you:

CodaKid

CodaKid is an online school that offers some seriously good coding courses for teens. It's not free, but it does have an online help desk if you have questions. They even run tech camps over the holidays if you'd prefer to learn to code with other people around you!

Alice

Alice is a desktop app that allows you to learn in 3D, so it is perfect if you already know the basics. You learn coding will making interactive games or animated videos that you can then share with your friends online.

App Inventor

App Inventor is made by MIT and is an excellent resource that uses drag-and-drop coding blocks. You can make your own app, and because it's cloud-based you can work on it from many devices over the summer. It is very beginner-friendly and focuses on using your creativity, plus you'll have a cool app to show your friends!

Vidcode

VidCode is a super exciting platform that uses existing social media culture to encourage you to code cool things. You'll learn to make videos, animations, and memes and can share them straight to Facebook or Instagram, and they also have really great support for if you get stuck.

It's worth trying a few different programs to see what works for you, but the great news is that by the end of the summer you'll be a confident coder... cool!

Week 4 - PSAT/NMSQT Prep

- Math
- Evidence Based Reading

https://www.lumoslearning.com/a/slh9-10

See Page 7 for Signup details

Weekly Fun Summer Photo Contest

Take a picture of your summer fun activity and share it on Twitter or Instagram

Use the **#SummerLearning** mention

@LumosLearning on Twitter or

@lumos.learning on Instagram

Tag friends and increase your chances of winning the contest

Participate and stand a chance to WIN $50 Amazon gift card!

Day 1

1. Eric is estimating the number of bacteria in his circular petri dish. The dish has a diameter of 10 cm. If there are 50 bacteria in every square cm, about how many bacteria are in the dish.

   Ⓐ 78.5
   Ⓑ 314
   Ⓒ 3,925
   Ⓓ 15,700

2. A beaker that weighs 3g when empty is completely filled with water. The beaker has a radius of 4 cm and is 12 cm high. The density of water is $\frac{1g}{cm^3}$. If the full beaker is placed on a scale what would the reading be?

   Ⓐ 602.88 g
   Ⓑ 605.88 g
   Ⓒ 610 g
   Ⓓ 704.59 g

3. A track is made in the shape of a perfect circle. The diameter of the track is 1 mile. How many complete laps will the cars have to make to have driven at least 600 miles?

   Ⓐ 190 laps
   Ⓑ 191 laps
   Ⓒ 192 laps
   Ⓓ 314 laps

4. A cone is placed inside a cylinder in such a way that the cross section shown below results. The radius of the cylinder and cone is 2 cm. The height of the cylinder is 10 cm and the height of the cone is 5 cm. What is the volume of the cylinder that is not filled by the cone?

   Ⓐ 345.4 cm³
   Ⓑ 104.67 cm³
   Ⓒ 245.7 cm³
   Ⓓ 156.8 cm³

5. Below is a dot plot showing the scores students received on a quiz. Which of the following statements is true based on the dot plot?

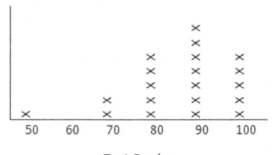

Test Grades

Ⓐ One student received a score of 60.
Ⓑ More students scored a 90 than any other score.
Ⓒ 70 was the score received by the fewest students.
Ⓓ More students scored a 70 or 80 than scored a 90.

Analyze U.S. Documents of Historical and Literary Texts

Day 1

Common Sense

Some writers have so been annoyed by society with the government as to leave little or no distinction between them whereas they are not only different but have different origins. Society is produced by our wants and the government by our wickedness. The former promotes our happiness positively by uniting our affections and the latter negatively by restraining our vices. The one encourages communication and the other creates differences. The first is a supporter, and the last is a punisher.

Society in every state is a blessing, but government even at its best state is but a necessary evil. In its worst state, it is an intolerable one. For when we suffer or are exposed to the same miseries by a government, our misfortune is heightened by reflecting that we furnish the means by which we suffer. Government is the badge of lost innocence, and the palaces of kings are built on the ruins of the places of paradise. If the impulses of a clear conscience, clear expectations, and irresistibly obeyed officials are there, then man would need no other lawgiver. However, that is not the case because he finds it necessary to surrender up a part of his property to furnish means for the protection of the rest. This, he is induced to do by the same cautiousness which in every other case advises him out of two evils to choose the least. Wherefore, security should be the true policy from the government, so that it follows whatever form appears most likely to ensure it to us with the least expense and greatest benefit is preferable to all others.

In order to gain a clear and just idea of the design of the government, let us suppose a small number of people settled in some sequestered part of the earth, unconnected with the rest, they will then represent the first peopling of any country or of the world. In this state of natural liberty, society will be their first thought. A thousand reasons will excite them thereto, the strength of one man is so unequal to his wants, and his mind so unfitted for continuous isolation, that he is soon obliged to seek assistance and relief of another. He requires the same. Four or five united people would be able to raise a tolerable dwelling in the midst of a wilderness, but one man might work out of the common period of life without accomplishing anything. When he had chopped for his timber, he could not remove it, nor build with it after it was removed. Therefore, hunger in the meantime would bother him from his work,and every different need calls him in a different way. Disease would disable him from living and reduce him to a state in which he might rather be said to perish than to die.

Thus, necessity would soon form our newly arrived emigrants into society the common blessings of the obligations of law and government unnecessary while they remained perfectly fair to each other. They prevail the first difficulties of emigration which bound them together in a common cause. They will begin to relax in their duty and attachment to each other. This is careless of establishing some form of government to supply the defect of moral virtue.

Some convenient tree will afford them a State-House, under the branches of which, the whole colony may assemble to deliberate on public matters. It is more than probable that their first laws will have the title only of regulations, and be enforced by no other penalty than public disfavor. In this first parliament every man, by natural right, will have a seat.

6. **Paine wrote "Common Sense" right before the American Revolutionary War when the United States was declaring its freedom from Great Britain.**
   Which answer best shows the significance of "Common Sense"?

   Ⓐ It focuses on the freedoms that Great Britain is not allowing the colonists.
   Ⓑ It compares and contrasts Great Britain's government with the one that should be used.
   Ⓒ It provides examples that the leaders are abusing their power.
   Ⓓ It declares that Great Britain's government is unjustified in their actions.

Declaration of Independence

The history of the present King of Great Britain is a history of repeated injuries and seizures, all having in direct object the establishment of an absolute Tyranny over these States. To prove this, let Facts be submitted to a candid world.

He has refused his Assent to Laws, the most wholesome and necessary for the public good.

He has forbidden his Governors to pass Laws of immediate and pressing importance unless suspended in their operation till his Assent should be obtained; and when so suspended, he has utterly neglected to attend to them.

He has refused to pass other Laws for the accommodation of large districts of people, unless those people would relinquish the right of Representation in the Legislature, a right inestimable to them and formidable to tyrants only.

He has called together legislative bodies at places unusual, uncomfortable, and distant from the depository of their Public Records, for the sole purpose of fatiguing them into compliance with his measures.

He has dissolved Representative Houses repeatedly, for opposing with manly firmness his invasions on the rights of the people.

He has refused for a long time, after such dissolutions, to cause others to be elected; whereby the Legislative Powers, incapable of Annihilation, have returned to the People at large for their exercise; the State remaining in the meantime exposed to all the dangers of invasion from without, and convulsions within.

He has endeavored to prevent the population of these States; for that purpose obstructing the Laws of Naturalization of Foreigners; refusing to pass others to encourage their migration hither, and raising the conditions of new Appropriations of Lands.

7. **America is about to go to war with Great Britain for their freedom. Why would this pamphlet cause issues with Great Britain?**

Ⓐ Because it would insult Great Britain who makes the laws for the colonists.
Ⓑ Because it would cause Great Britain to stop trading with the colonists.
Ⓒ Because it would alert Great Britain that the United States is about to declare war.
Ⓓ Because it would create a panic among the British and the colonists.

8. **What theme is addressed in the "Declaration of Independence"?**

9. **In what way does Jefferson address the problems with the British government?**

Ⓐ lightheartedly
Ⓑ condescending
Ⓒ elusive
Ⓓ directly

10. What is the significance of this part of the Declaration of Independence?

Day 2

1. **What is the difference between the means of these two sets of data?**
   Set A: {2,4,6,8,10,12}
   Set B: {3,5,7,9,11,13}

   Ⓐ 7
   Ⓑ 1
   Ⓒ 8
   Ⓓ 15

2. **What is the difference in the medians of Set A and Set B, displayed in the line plot below?**

   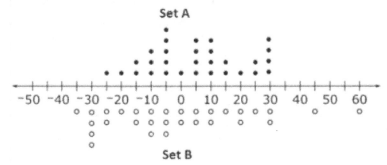

   Ⓐ 5
   Ⓑ 10
   Ⓒ -5
   Ⓓ 0

3. **What effect does a group of very small values have on the mean and median of a data set?**

   Ⓐ The mean and median both decrease
   Ⓑ The mean is not changed, but the median is decreased
   Ⓒ The mean and median are not changed
   Ⓓ The mean is decreased, but the median is increased

4. **What is a uniform data set look like?**

   Ⓐ The values are arranged from smallest to largest
   Ⓑ The values are all the same
   Ⓒ The values are arranged from largest to smallest
   Ⓓ The values are uniformly distributed between the smallest and largest

5. Given a set of data that is normally distributed and has a mean of 90.12 and a standard deviation of 4.3, find the range in which 95% of data lies?

  Ⓐ Above 85.82 and below 94.42
  Ⓑ Below 85.82 and above 94.42
  Ⓒ Above 98.72 and below 81.52
  Ⓓ Below 98.72 and above 81.52

Grammar and Usage Conventions

Day 2

6. "Jim's brand new sports car skidded to a stop in the driveway."
   In the quote above, which words represent a verb phrase?

  Ⓐ Jim's brand new sports car
  Ⓑ Skidded to a stop
  Ⓒ In the driveway
  Ⓓ None of the above

7. "Jim's brand new sports car skidded to a stop in the driveway."
   In the quote above, which words represent a noun phrase?

  Ⓐ Jim's brand new sports car
  Ⓑ Skidded to a stop
  Ⓒ In the driveway
  Ⓓ None of the above

8. "Jim's brand new sports car skidded to a stop in the driveway."
   In the quote above, which words represent a prepositional phrase?

  Ⓐ Jim's brand new sports car
  Ⓑ Skidded to a stop
  Ⓒ In the driveway
  Ⓓ None of the above

9. Which of the following is a sentence fragment?

  Ⓐ Xiomara was late for class.
  Ⓑ When Jared went home.
  Ⓒ After Rachel took her test, she drew a picture.
  Ⓓ Because of the thunderstorm, the game was canceled.

10. Identify the problem with this sentence: "Smith forgot to feed the dog, his mom was disappointed."

Ⓐ comma splice
Ⓑ run on
Ⓒ sentence fragment
Ⓓ faulty parallelism

Challenge Yourself!

- **Interpreting Categorical & Quantitative Data**
- **Grammar and Usage Conventions**

https://www.lumoslearning.com/a/dc9-22

Day 2

See Page 7 for Signup details

1. Billy has a bag that contains only blue and green marbles. If he has 30 marbles and 3 are green, what is the probability he will choose a blue marble at random?

    Ⓐ 10%
    Ⓑ 20%
    Ⓒ 80%
    Ⓓ 90%

2. Mr. Smith teaches a Spanish class to 6th, 7th, and 8th graders. Below is a chart showing the grade level of students and the letter grade they earned in the class. Mr. Smith needs to select a 6th grader with an A in his class. What is the probability he would select such a student if he selects in a completely random fashion from all his students?

Grade	A	B	C
6th	5	6	7
7th	4	8	8
8th	9	2	1

    Ⓐ 10%
    Ⓑ 27.78%
    Ⓒ 36%
    Ⓓ 52.34%

3. Given the scatter plot below, what type of function expresses the correlation between the two variables?

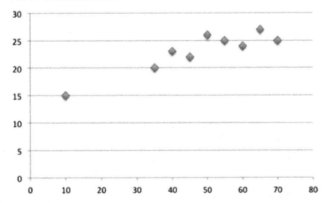

    Ⓐ Linear
    Ⓑ Exponential
    Ⓒ Quadratic
    Ⓓ Polar

4. **Dr. Winthrop is conducting a study to test her hypothesis that the number of mosquitos in an area is dependent on the temperature. She believes the correlation is described by y=3x+4.5. What does x represent in this equation?**

    Ⓐ The number of mosquitos in an area
    Ⓑ The temperature
    Ⓒ The total number of mosquitos counted throughout the study
    Ⓓ A coefficient that creates a quadratic formula

5. **What is the slope(m) and y-intercept(b) of the equation below?**
   y=x

    Ⓐ m=1 b=1
    Ⓑ m=0 b=1
    Ⓒ m=1 b=0
    Ⓓ m=0 b=0

Capitalization, Punctuation, and Spelling

Day 3

6. **Which of the following does not need to be capitalized in a sentence?**

    Ⓐ first letter in the sentence
    Ⓑ proper nouns
    Ⓒ common nouns
    Ⓓ acronyms

7. **Correct the run-on: "The company's expenses went over budget in February spending was reduced in March."**

    Ⓐ The company's expenses went over budget in February, spending was reduced in March.
    Ⓑ The company's expenses went over budget in February; therefore, spending was reduced in March.
    Ⓒ The company's expenses went over budget in February: spending was reduced in March.
    Ⓓ No change.

8. **How should you correct the following sentence?**
   Your one of the top performers on the team, which is why you're contribution is so important.

9. Name the type of error in the following sentence:
   "The whether outside was getting damp and muggy."

   Ⓐ capitalization
   Ⓑ punctuation
   Ⓒ homophone usage
   Ⓓ sentence fragment

10. A _____ is when two independent clauses are combined by a comma without a coordinating conjunction.

    Ⓐ sentence fragment
    Ⓑ run on
    Ⓒ compound sentence
    Ⓓ comma splice

Conditional Probability & the Rules of Probability

1. Mark was sorting his tools after a long project and gathered his 14 screwdrivers (S) his 4 hammers (H) and his 8 pairs of pliers (L) and placed them in a drawer (U) Which Venn diagram correctly shades the open space in the drawer after Mark finished putting away his tools?

Ⓐ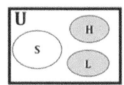

Ⓒ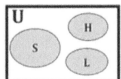

Ⓑ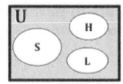

Ⓓ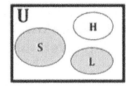

2. In December, it snowed (S) days, it rained (R) days, it was sunny (Y) days, and it was cloudy (C) days. Which region correctly shades the days of the month when it either snowed or rained?

Ⓐ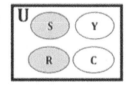

Ⓒ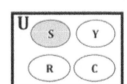

Ⓑ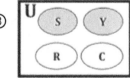

Ⓓ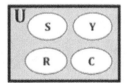

3. A standard deck of playing cards has 52 cards, with 13 hearts, 13 clubs, 13 spades, and 13 diamonds. Suppose you randomly pick a card and it is a diamond, but you do not replace the card back into the deck. What is the probability of randomly picking another diamond, and are these two events independent or dependent?

Ⓐ $\frac{12}{51}$ - independent

Ⓑ $\frac{12}{51}$ - dependent

Ⓒ $\frac{1}{4}$ - dependent

Ⓓ $\frac{1}{4}$ - independent

4. A standard deck of playing cards has 52 cards, with 13 hearts, 13 clubs, 13 spades, and 13 diamonds. Each suit has one jack, one queen, one king, and one ace, in addition to the numbered cards. Suppose you randomly pick a card and it is an ace, but you do not replace the card back into the deck. What is the probability of randomly picking another ace, and are these two events independent or dependent?

Ⓐ $\frac{1}{17}$ - independent

Ⓑ $\frac{1}{17}$ - dependent

Ⓒ $\frac{1}{13}$ - dependent

Ⓓ $\frac{1}{13}$ - independent

5. A survey of all of the members of an outdoor youth camp asked the camp attendees to identify which outdoor activity they most preferred to participate in next week. The result of the survey is in the table below. What is the probability that a randomly selected attendee is interested in hiking given that the student is a girl?

Ⓐ $\frac{1}{5}$

Ⓑ $\frac{10}{23}$

Ⓒ $\frac{23}{50}$

Ⓓ $\frac{50}{113}$

	Hiking	Kayaking	Total
Boys	65	72	137
Girls	50	63	113
Total	115	135	250

Functions of Language

Day 4

6. Select the title of the source in this Works Cited entry for an article. Beck, Julie, "The Psychology of Voldemort", The Atlantic, 23 Sept. 2015.

Ⓐ Julie Beck
Ⓑ The Psychology of Voldemort
Ⓒ The Atlantic
Ⓓ September 23, 2015

7. Select the author of the source in this Works Cited entry for an article. Beck, Julie. "The Psychology of Voldemort." The Atlantic, 23 Sept. 2015.

Ⓐ Julie Beck
Ⓑ The Psychology of Voldemort
Ⓒ The Atlantic
Ⓓ Beck, Julie

8. Complete the following analogy: Trustworthy is to credible as faulty is to _____.

Ⓐ strong
Ⓑ fallible
Ⓒ reliable
Ⓓ conclusive

9. Select the author of the source in this Works Cited entry for a book. Snow, Shane, Smartcuts, New York, Harper Business 2014.

Ⓐ Shane Snow
Ⓑ Smartcuts
Ⓒ New York
Ⓓ Harper Business

10. Select the title of the source in this Works Cited entry for a book. Snow, Shane. Smartcuts. New York: Harper Business, 2014.

Ⓐ Shane Snow
Ⓑ Smartcuts
Ⓒ New York
Ⓓ Harper Business

1. **Which of the following is an example of categorical data?**

   Ⓐ Maximum miles per hour a car can reach
   Ⓑ Average miles / gallon of a car
   Ⓒ Color of a car
   Ⓓ Prices of different brands of cars

2. **A college lecture has 40 male students and 50 female students. What is the probability that a student chosen at random will be a male?**

   Ⓐ 55.56%
   Ⓑ 44.44%
   Ⓒ 50%
   Ⓓ 60%

3. **What four labels could you use in a two-way frequency table to represent the following data?**
   Is there a relationship between people who are overweight and people who snore at night?

   Ⓐ People who snore at night, people who don't snore at night, people who are overweight, people who are not overweight
   Ⓑ People who don't sleep well, people who do sleep well, people who are overweight, people who are not overweight
   Ⓒ People who like pillows, people who don't like pillows, people who are overweight, people who are not overweight
   Ⓓ None of these

4. **Given the following two-way frequency table groupings, what would you need to calculate to find the probability that person prefers Peas.**
   Group 1: Peas (P), Carrots (C), Green beans (G)
   Group 2: Drives the speed limit (L), Speeds (S)

   Ⓐ P(G)
   Ⓑ P(C)
   Ⓒ P(P)
   Ⓓ None of these

5. Given the following two-way frequency table groupings, what would you need to calculate to find the probability that person prefers Carrots given that they Speed.
   Group 1: Peas (P), Carrots (C), Green beans (G)
   Group 2: Drives the speed limit (L), Speeds (S)

   Ⓐ P(S|C)
   Ⓑ P(C|S)
   Ⓒ P(C)+P(S)
   Ⓓ None of these

Functions of Language (Contd.)

Day 5

6. Select the author's name in this Works Cited entry for a book.

   Foster, Thomas C. How to read literature like a professor: a lively and entertaining guide to reading between the lines. Harper, an imprint of Harper Collins Publishers, 2017.

   Ⓐ Harper Collins
   Ⓑ How to Read Literature Like a Professor
   Ⓒ Thomas C. Foster
   Ⓓ Foster C. Thomas

7. Select the title of the source in this Works Cited entry for an article.

   Coscarelli, Joe. "Taylor Swift's 'Reputation' Sells 1.2 Million Copies in Its First Week." The New York Times, 21 Nov. 2017.

   Ⓐ Joe Coscarelli
   Ⓑ Taylor Swift
   Ⓒ Reputation
   Ⓓ The New York Times

8. Which is the only piece of information a reader CAN NOT learn from this Works Cited entry?

   Shane, A. L. "Cell phones take parents' attention away from kids on playgrounds." Journal of the American Academy of Pediatrics, vol. 26, no. 1, Jan. 2011, pp. 5–15.

   Ⓐ The length of the article in pages
   Ⓑ The source of the article
   Ⓒ The date the reader downloaded the article
   Ⓓ The date the author published the article

9. **Complete the analogy. Graceful is to awkward as excruciating is to**

Ⓐ unbearable
Ⓑ mild
Ⓒ painful
Ⓓ excess

10. **Complete the analogy. Gasoline is to tank as money is to**

Ⓐ vault
Ⓑ silo
Ⓒ mattress
Ⓓ store

Challenge Yourself!

- **Interpreting Categorical & Quantitative Data**
- **Functions of Language(contd.)**

https://www.lumoslearning.com/a/dc9-25

See Page 7 for Signup details

Day 5

How to Make Extra Cash on Your Summer Break

As a teenager, you have a lot going on: trying to balance school, social life, and potentially a job. It can all be a lot to handle, and of course, study comes first. The summer break, however, is a great chance to use your free time to make some extra bucks. Whether you are saving for college, for a trip after graduation, or just for a bit of financial independence, here are our top tips to making money as a high school student both on and off the internet:

Offline

- **Babysitting** is the classic teen job, and there's a reason! The jobs are usually very flexible and local, and if you're lucky you'll have some fun! Once you've proven yourself trustworthy, it is likely that your name will get passed around to other families in the area, and before you know it you will be the most south-after sitter in the neighbourhood.

- If you're keen to get sweaty, **mowing lawns** is another service that can soon turn into a mini business. Create flyers to put in neighbours' mailboxes as a way of advertising.

- Another great gig in the great outdoors is **dog-walking**, and the advantage of this is you can usually walk a few dogs at once (with their owners' permission). You are most likely to get hired on a recurring schedule, and also may have inquiries for **pet-sitting** (after all, cats don't love going on walks!)

Online

- If you have any creative passions, there's a chance you could make money **selling on Etsy**. We all remember the slime craze, and that's a great example of tapping into current trends. Other ideas are tie-dye clothes, jewelery, or artwork.

- For a bit of extra cash you could always **take surveys** online. Sure, you aren't guaranteed huge amounts of money, but it's better than nothing!

- **Starting a blog** is one of the best ways to make money as a teen, and summer break is a perfect time to work on it. It will take time to naturally build a following, but as long as you can create shareable content, you could be making a lot in affiliate or ad money.

Any way you earn money, the key is to work hard and create good contacts... and hopefully have a bit of fun along the way!

Week 5 - PSAT/NMSQT Prep

- Math
- Evidence Based Writing & Language

https://www.lumoslearning.com/a/slh9-10

See Page 7 for Signup details

Weekly Fun Summer Photo Contest

Take a picture of your summer fun activity and share it on Twitter or Instagram

Use the **#SummerLearning** mention

@LumosLearning on Twitter or

@lumos.learning on Instagram

Tag friends and increase your chances of winning the contest

Participate and stand a chance to WIN $50 Amazon gift card!

Week 6 Summer Practice

Conditional Probability & the Rules of Probability (Contd.)

Day 1

Refer the below Venn diagram for question no. 1 & 2.

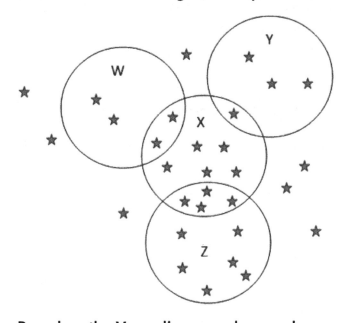

1. Based on the Venn diagram above, where each dot represents a possible outcome, what is P(X|Y)? Give your answer as a fraction.

2. Using the Venn diagram, find P(W|X).

   Ⓐ $\frac{2}{7}$ - independent

   Ⓑ $\frac{1}{6}$ - dependent

   Ⓒ $\frac{1}{3}$ - dependent

   Ⓓ None of these

Refer the below Venn diagram for question no. 3 & 4.

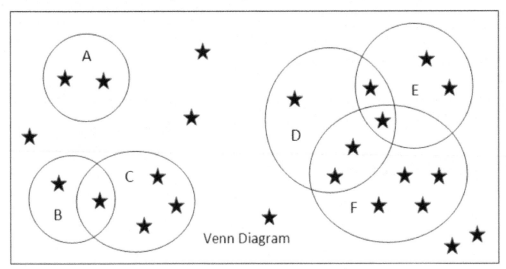

Venn Diagram

3. Based on the Venn diagram above, where each star represents a possible outcome, what is P(B or C)? Give your answer as a fraction.

4. Using just the Venn diagram above, find P(E or F).

    (A) $\frac{11}{24}$ - independent

    (B) $\frac{9}{24}$ - dependent

    (C) $\frac{5}{12}$ - dependent

    (D) None of these

5. Use the permutation formula to determine the number of outcomes for $_6P_2$.

Determining Unknown Words

6. Overconfident, overburdened, and overjoyed all begin with the prefix over-. What is the meaning of the prefix over-?

    (A) beneath
    (B) excessively
    (C) around
    (D) above

7. Which word is a synonym for accentuate?

    (A) decorate
    (B) hide
    (C) reflect
    (D) highlight

8. Which prefix means against?

    (A) mis-
    (B) anti-
    (C) non-
    (D) un-

9. Which root word means sound?

    (A) alter
    (B) carn
    (C) aud
    (D) arch

10. Which suffix means belief?

    Ⓐ -er
    Ⓑ -ness
    Ⓒ -tion
    Ⓓ -ism

Challenge Yourself!

- Conditional Probability & the Rules of Probability
- Determining Unknown Words

https://www.lumoslearning.com/a/dc9-26

Day 1

See Page 7 for Signup details

Using Probability to Make Decisions

1. **How many outcomes would be possible for $_5C_5$?**

    Ⓐ 120 outcomes
    Ⓑ 24 outcomes
    Ⓒ 1 outcome
    Ⓓ None of these

2. **A certain health insurance policy covers 80% of all health related costs. Studies show that it is about 40% likely that you will break your arm and 20% likely that you will break your leg in the next year. A broken leg costs on average $3500 and a broken arm costs on average $2000 after insurance has paid their 80%. What is the difference in expected value if you choose not to buy the insurance over buying it?**

    Ⓐ $7,500
    Ⓑ $6,000
    Ⓒ $1,500
    Ⓓ $9,000

3. **As of 2017, the richest people in the world and their net worth are listed in the table below in billions. If you randomly selected someone from this list, what would expect his or her net worth to be?**

    Ⓐ 68
    Ⓑ 70
    Ⓒ 69
    Ⓓ 67

Billionaire	Worth (in billions)
Jeff Bezos	90.6
Bill Gates	90
Amancio Ortega	83.2
Warren Buffett	74.3
Mark Zuckerberg	72.2
Carlos Slim Helu	69.7
Larry Ellison	62.1
Michael Bloomberg	53.3
Bernard Arnault	53
Charles Koch	48.5
David Koch	48.5

4. The chances of you winning a certain lottery game are 1 in 25. What are the chances you'll win 2 or fewer times if you buy 30 tickets?

5. If you are taking a test with multiple choice questions, what is the probability that you will randomly answer 2 questions with 4 answer choices each and get the first one wrong and the second one right?

Ⓐ 0.375
Ⓑ 0.25
Ⓒ 0.1875
Ⓓ 0.75

Figurative Language

Day 2

6. Which word has the more positive connotation? Choose the correct answer.

Ⓐ inexpensive
Ⓑ cheap

7. Which word has the more negative connotation? Choose the correct answer.

Ⓐ inexpensive
Ⓑ cheap

8. "Old news" is an example of which type of figurative language?

Ⓐ metaphor
Ⓑ simile
Ⓒ onomatopoeia
Ⓓ oxymoron

9. The rain sounded like small pellets falling onto the roof.

   The sentence above uses which type of figurative language?

   Ⓐ metaphor
   Ⓑ simile
   Ⓒ hyperbole
   Ⓓ personification

10. The clouds sobbed, drenching the earth with their tears. The sentence above uses which type of figurative language?

   Ⓐ metaphor
   Ⓑ simile
   Ⓒ hyperbole
   Ⓓ personification

Challenge Yourself!

• **Using Probability to Make Decisions**

• **Figurative Language**

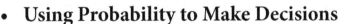

https://www.lumoslearning.com/a/dc9-27

Day 2

See Page 7 for Signup details

1. A statistician is working for Sweet Shop USA and has been given the task to find out what the probability is that the fudge machine malfunctions messing up a whole batch of fudge in the process. Each malfunction of the machine costs the company $250. The statistician calculates the probability is 1 in 20 batches of fudge will be lost due to machine malfunction. What is the expected value of this loss for one month if the company produces 20 batches of fudge each day?

   Ⓐ $3750
   Ⓑ $150,000
   Ⓒ $7500
   Ⓓ $375

2. If you bet $1 on a hockey game where your team has a .70 chance of winning with a $2 payout for winning, how much money would you expect to make after betting on 20 games?

   Ⓐ $28
   Ⓑ $6
   Ⓒ $22
   Ⓓ $40

3. You buy a trick deck of cards from one of those gag gift stores in the mall. The deck has 20 cards, 10 that are hearts, 5 that are diamonds, 3 that are spades and 2 clubs. You take them to your family reunion and tell your cousin (who has always picked on you) that it will cost him $2 per game to play with you. If he randomly picks a heart he wins nothing, for a diamond he wins $1, for a spade $3, and for a club $5. What is your cousin's expected value after 10 rounds of playing this game?

   Ⓐ $8
   Ⓑ -$8
   Ⓒ $12
   Ⓓ -$20

4. Items that are used in a fair decision making process should be checked for what?

   Ⓐ Different shapes
   Ⓑ Different sizes
   Ⓒ Different weights
   Ⓓ All of these

5. **Which decision making tool would be considered fair?**

   Ⓐ Tossing a fair die
   Ⓑ Using a random number generator
   Ⓒ Both A and B
   Ⓓ None of these

Academic and Domain-Specific Vocabulary

Day 3

6. **Select the word that best completes the sentence given below:**

   "The football team we are playing this Friday hasn't lost a game yet, making them a(n) _____ opponent."

   Ⓐ acceptable
   Ⓑ formidable
   Ⓒ credible
   Ⓓ reasonable

7. **Select the word that best completes the sentence given below:**

   "When Sara felt alone, she struggled to feel better about herself, but when she began to make friends at school, her personality began to _____."

   Ⓐ exult
   Ⓑ maneuver
   Ⓒ flourish
   Ⓓ descend

8. **Select the word that best completes the sentence given below:**

   "Chad didn't feel like celebrating when he woke up this morning, but when all his friends surprised him with a party for his birthday, he felt more _____."

   Ⓐ jovial
   Ⓑ melancholy
   Ⓒ laconic
   Ⓓ random

9. **Select the word that best completes the sentence given below:**

"The citywide vote last year to increase taxes and to increase teacher pay was divided, but the vote was _____ this year, which will be good news for teachers."

Ⓐ voracious
Ⓑ receded
Ⓒ reposed
Ⓓ unanimous

10. **Select the word that best completes the sentence given below:**

"My appetite was _____ after going without food for two days."

Ⓐ voracious
Ⓑ quenched
Ⓒ tumultuous
Ⓓ wretched

Challenge Yourself!

- **Using Probability to Make Decisions**
- **Academic and Domain-Specific Vocabulary**

https://www.lumoslearning.com/a/dc9-28

Day 3

See Page 7 for Signup details

Creating Equations

1. McDonald's has come up with a new menu item that they think will appeal to 30% of their customers. The executives have decided that the chances that the item will be bought by a random person or not be bought by a random person are 50%. Are they correct or incorrect in their assumption?

2. Ms. Baker is making cookies for her family's holiday party. Her favorite recipe calls for 2 cups of flour and $\frac{1}{2}$ cup of chocolate chips. She accidentally put in 3 cups of flour, how many total chocolate chips will she need to ensure the cookies come out correctly?

   Ⓐ $\frac{3}{4}$ cup

   Ⓑ 1 cup

   Ⓒ $\frac{1}{3}$ cup

   Ⓓ $\frac{1}{8}$ cup

3. The width of a rectangular classroom is 5 less than twice its length. If the area of the classroom is 50 square meters, what equation can be used to find the length in meters?

   Ⓐ $\frac{50}{l} = 2l - 5$

   Ⓑ $\frac{50}{l} = 5 - 2l$

   Ⓒ $l(2l - 5) = 50$

   Ⓓ Both A and C

4. A hot air balloon leaves the ground and rises at a steady rate of 3 kilometers (k) per minute (m). Which equation below represents this relationship?

   Ⓐ k = 3 + m
   Ⓑ k = 3m
   Ⓒ m = 3k
   Ⓓ m = k + 3

5. Annie's contractor put down carpet in her new office at a rate of 4 square feet per hour. After 2 hours, he had 40 square feet left. Which function, C(t) where t is time, shows a formula for this situation?

Ⓐ $C(t) = -4t + 48$
Ⓑ $C(t) = 4t + 40$
Ⓒ $C(t) = 4t + 48$
Ⓓ $C(t) = 4t - 40$

Day 4

Write Arguments to Support Claims

6. Which of the following is the best thesis statement for an argumentative essay?

Ⓐ Gap years are becoming more and more popular among high school graduates.
Ⓑ This essay is going to be about the benefits of gap years.
Ⓒ Although high school graduates have traditionally entered college directly after graduation, there are a number of reasons to consider taking a gap year instead.
Ⓓ Gap years are helpful.

7. An argumentative thesis statement must do all of the following EXCEPT?

Ⓐ Make a claim
Ⓑ Be debatable
Ⓒ Be an opinion
Ⓓ State a fact

8. The part of your argument that considers what the opposition has to say is called?

Ⓐ call to action
Ⓑ counterclaim
Ⓒ thesis statement
Ⓓ claim

9. Which of the following should you include in the conclusion of your argument?

Ⓐ Evidence
Ⓑ Call to action
Ⓒ Counterclaim
Ⓓ Claim

10. Which of the following maintains a formal style and objective tone?

Ⓐ The folks enrolled at the local university helped them do some research.
Ⓑ The students of the prestigious institution participated in a unique research study.
Ⓒ The kids at the best college in town were part of an experiment.
Ⓓ The college kids helped with the research.

Challenge Yourself!

- **Creating Equations**
- **Write arguments to support claims**

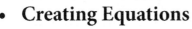

https://www.lumoslearning.com/a/dc9-29

See Page 7 for Signup details

Day 4

Day 5

1. The ratio of staff to guests at a gala was 3 to 5. If there were a total of 576 people in the ballroom, how many guests were at the gala?

   Ⓐ 216
   Ⓑ 360
   Ⓒ 300
   Ⓓ 276

2. James decided to invest in some lawn mowing equipment to start his own lawn mowing business. He spent $875 on a lawn mower and it costs him $20 in gasoline every day. If he earns $35 per day for mowing lawn, how many days must he work to break even?

   Ⓐ 60
   Ⓑ 59
   Ⓒ 58
   Ⓓ 61

3. Verizon wireless charges a set fee of $55, for it's basic plan which includes several 3 gigabytes of data. If a customer goes over their data plan and uses more than the 3 gigabytes, they are charged an additional $40. If V represents the cost of and d represents the total number of gigabytes of data, which equation would represent the bill for a customer who went over their plan?

   Ⓐ $V = 55 + 40(d - 3)$
   Ⓑ $V = 55 + 40(3 - d)$
   Ⓒ $V = 40 + 55(d - 3)$
   Ⓓ $V = 40 + 55(3 - d)$

4. The formula for the area of a triangle is $A = \frac{1}{2}bh$. Solve for b.

   Ⓐ $b = \frac{A}{2h}$

   Ⓑ $b = \frac{2A}{h}$

   Ⓒ $b = 2Ah$

   Ⓓ $b = \frac{Ah}{2}$

5. The volume of a cylinder can be expressed as $V = \pi r^2 h$. Solve for h.

Ⓐ $h = \dfrac{V}{\pi\sqrt{r}}$

Ⓑ $h = Vr^2$

Ⓒ $h = \dfrac{V}{\pi r^2}$

Ⓓ $h = V - \pi r^2$

Write Arguments to Support Claims (Contd.)

Day 5

6. The most important sentence(s) in your argument is _____

7. What are five steps in the writing process that you should use to prepare your argument?

8. Which of the following is part of the perfect paragraph:

Ⓐ Topic Sentence
Ⓑ Supporting Sentences
Ⓒ Clincher
Ⓓ All of the above
Ⓔ None of the above

9. In a traditional 5 paragraph argument essay, how many points should you discuss in your thesis?

Ⓐ One
Ⓑ Seven
Ⓒ Three
Ⓓ Two

10. Which paragraph includes the thesis in a 5 paragraph essay?

Ⓐ Supporting Paragraph #1
Ⓑ Conclusion
Ⓒ Introduction
Ⓓ None of the above

Challenge Yourself!

- **Creating Equations**
- **Write arguments to support claims (contd.)**

  https://www.lumoslearning.com/a/dc9-30

Day 5

See Page 7 for Signup details

Where Do I Intern Before Starting College?

Everyone tells you how important it is to get an internship so you can gain experience, but how do you go about getting one? Sometimes it can be about who you know, but it can also be about how you present yourself. That is why it is important to build a portfolio before you start applying for internships. A few other key factors to getting a great internship include networking, attending job fairs, and always searching for new listings.

Here are some steps to follow to get a great internship that will help you to advance toward your goals.

1. **Start your search early:** Deadlines may be closer than they appear. Many companies like to hire interns in November and have them work over the winter months. Getting applications out early gives you time to make some connections in the field.

2. **Identify your career goals:** You can start with broad goals and narrow them down later. Allowing yourself the opportunity to explore some options is advantageous if you have not settled on a specific career path. An internship will allow you to gain exposure to various careers.

3. **Network:** This is where you use who you know to meet more people. Talk to everyone you know about your interests and objectives and see who they may connect you with. You can grow your network through family, friends, professors, and career counselors. If you do not have an account on LinkedIn, you may want to create one. It is a great place to build a community of like-minded individuals and connect with others who share interests and goals.

4. **Use online resources and searches:** Do all the searching you can to find out about companies that are looking for interns.

5. **Go to job fairs:** Watch for notices in your community about job and career fairs. Many employers will use career fairs as a way of hiring interns and full-time employees. When you attend, take some copies of your resume, your portfolio, and a prepared 60-second introduction that will inform potential employers of the value you feel you can contribute to the company. If you speak with potential employers, make sure to follow up with them following the fair.

6. **Contact potential employers:** Even if a company is not advertising for an intern, that does not mean they would not consider hiring one. Visit companies you have an interest in and ask about any possible opportunities for a summer job or an internship.

Following the steps above is sure to land you an internship where you will be able to grow, learn, and start working toward your career goals.

Week 6 - PSAT/NMSQT Prep

- Math
- Evidence Based Writing & Language

https://www.lumoslearning.com/a/slh9-10

See Page 7 for Signup details

Weekly Fun Summer Photo Contest

Take a picture of your summer fun activity and share it on Twitter or Instagram

Use the **#SummerLearning** mention

@LumosLearning on Twitter or

@lumos.learning on Instagram

Tag friends and increase your chances of winning the contest

Participate and stand a chance to WIN $50 Amazon gift card!

Week 7 Summer Practice

Reasoning with Equations & Inequalities

1. Solve for x: $2x-4=3x-11$.

    (A) $x=15$
    (B) $x=7$
    (C) $x=-15$
    (D) $x=-7$

2. What is the multiplicative inverse of $\frac{2}{3}$?

    (A) -1

    (B) $-\frac{3}{2}$

    (C) $-\frac{2}{3}$

    (D) $\frac{3}{2}$

3. In the set of positive integers, what is the solution set of the inequality $2x -3<5$?

    (A) $\{0,1,2,3\}$
    (B) $\{1,2,3\}$
    (C) $\{0,1,2,3,4\}$
    (D) $\{1,2,3,4\}$

4. Which table of values represents the quadratic $y=x^2-6x+8$?

    (A)

X	Y
-3	3
-4	0
-3	-1
-2	0

    (C)

X	Y
1	3
2	1
3	-1
4	1

    (B)

X	Y
1	3
2	0
3	-1
4	0

    (D)

X	Y
0	3
1	0
2	-1
3	0

5. Solve: $4x^2 - 8x = 0$.

Ⓐ 4, −8
Ⓑ 0, −2
Ⓒ 0, 2
Ⓓ −4, 8

Write informative/explanatory texts

Day 1

6. **Which of the following would be an example of informational writing?**

Ⓐ A story about an evil queen
Ⓑ A newspaper report about a local fair
Ⓒ A presentation to convince the school board to change the school start time
Ⓓ A poem about a struggling friendship

7. **What pattern of organization does the author use in the passage below?**

Jayden wondered why he wasn't able to stay awake during class recently. The more he thought about it, the more he realized he was probably spending too much time playing his video games at night before going to bed.

Ⓐ Problem and solution
Ⓑ Compare and contrast
Ⓒ Cause and effect
Ⓓ Order of importance

8. **What pattern of organization does the author use in the passage below?**

Trash build-up in the local parks is a serious problem. Local service groups, however, can make a difference by just spending an hour each week picking up trash.

Ⓐ Problem and solution
Ⓑ Compare and contrast
Ⓒ Cause and effect
Ⓓ Order of importance

9. **What pattern of organization does the author use in the passage below?**

Writing an essay isn't that hard if you follow a process. First, you should brainstorm some ideas until you find something you're confident you can write well about. Then, you can develop a clear controlling idea for your essay. Follow this with an outline of your thoughts on the controlling idea, then write your first draft. Finally, you'll revise and edit your draft until it's ready for submission.

Ⓐ Problem and solution
Ⓑ Compare and contrast
Ⓒ Cause and effect
Ⓓ Order of importance

10. **What pattern of organization does the author use in the passage below?**

Desktop computers and laptop computers both perform a lot of the same functions. For example, you can pull up a word processing program on either device and type up a report or send an email. Laptops, however, are often preferred by busy people who are always on the go because they can take their computer with them and work from anywhere. Desktop computers, though they often have more power and space, don't provide the convenience of laptop computers.

Ⓐ Problem and solution
Ⓑ Compare and contrast
Ⓒ Cause and effect
Ⓓ Order of importance

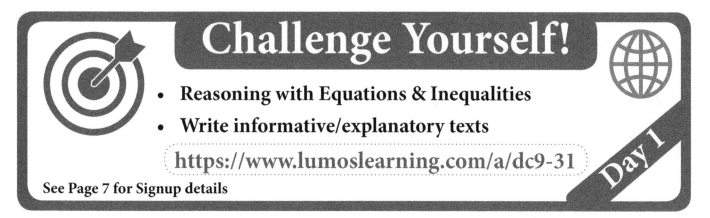

Challenge Yourself!

- **Reasoning with Equations & Inequalities**
- **Write informative/explanatory texts**

https://www.lumoslearning.com/a/dc9-31

Day 1

See Page 7 for Signup details

1. Use the elimination method to solve the system of equations.

$$\begin{cases} 5x+2y=0 \\ 3x-2y=-16 \end{cases}$$

Ⓐ $(-5, 2)$
Ⓑ $(2, -5)$
Ⓒ $(-2, 5)$
Ⓓ $(5, -2)$

2. What is the solution to the system $\begin{cases} -2x=y-1 \\ y+x=4 \end{cases}$

Ⓐ $(2, -4)$
Ⓑ $(7, -3)$
Ⓒ No solution
Ⓓ $(-3, 7)$

3. Kim ran 4 miles farther than Alex. They ran a combined distance of 20 miles. How far did each of them run?

Ⓐ Kim 10 miles, Alex 10 miles
Ⓑ Kim 11 miles, Alex 9 miles
Ⓒ Kim 12 miles, Alex 8 miles
Ⓓ No solution

4. Two cans of paint and one roller cost $62. Five cans of the same paint and two rollers cost $151. Find the cost of one can of paint and one roller.

Ⓐ One can of paint $27, one roller $8
Ⓑ One can of paint $8, one roller $27
Ⓒ One can of paint $25, one roller $12
Ⓓ One can of paint $30, one roller $2

5. Would this system of equations intersect? If so, how many times?

$$\begin{cases} y=x^2+5x+9 \\ y=\frac{1}{4}x-8 \end{cases}$$

Ⓐ Yes, one
Ⓑ No
Ⓒ Yes, two
Ⓓ Yes, three

Day 2

6. The character the reader is most meant to identify within a story is known as the:

- Ⓐ foil
- Ⓑ protagonist
- Ⓒ antagonist
- Ⓓ sidekick

7. What point of view does the author use in the passage below?

Jack wondered if Sarah knew what secrets he held from his childhood. She didn't seem to treat him any differently, but still he wondered.

- Ⓐ Third person objective
- Ⓑ First person
- Ⓒ Third person omniscient
- Ⓓ Third person limited

8. The part of a story that establishes the setting and the characters is known as:

- Ⓐ the exposition
- Ⓑ the rising action
- Ⓒ the climax
- Ⓓ the resolution

9. When a narrative is interrupted by a scene that occurred earlier in time, the scene is known as a/an:

- Ⓐ foreshadowing
- Ⓑ rising action
- Ⓒ flashback
- Ⓓ exposition

10. Match the pronoun to the point of view.

	First Person	Second Person	Third Person
I, We	○	○	○
Me, Us	○	○	○
He, she, it	○	○	○
You, yours	○	○	○

Challenge Yourself!

- **Reasoning with Equations & Inequalities**
- **Write Narratives**

https://www.lumoslearning.com/a/dc9-32

See Page 7 for Signup details

Day 2

Day 3

1. Karen was trying to decide the y-value when $x=-6$ for the equation $y=4-9x$. She arrived at a y-value of -50 but that is wrong. Below is her work. Where did she make a mistake and how could you fix it?

$y=4-9x$
$y=4-9(-6)$
$y=4-54$
$y=-50$

2. Solve the equations in each row below for $x=-2$ and select all the correct answers.

	13	-21	-12
$y=8x-5$	○	○	○
$y=-3x+7$	○	○	○
$y=-x-14$	○	○	○

3. Find the intersection point of $y=\log(2x)$ and $y=x+3$.

Ⓐ They do not intersect
Ⓑ (2, 3)
Ⓒ (0, 0)
Ⓓ (0, 3)

4. Which inequalities would have the following points as part of their solution? Check all that apply.

	(0, 2)	(1, 7)	(5, 1)
$y_1>3x-2$	☐	☐	☐
$y_2\leq -x^2+5$	☐	☐	☐
$y_4\leq \frac{-1}{4}(x-5)^2+4$	☐	☐	☐
$y_3>-x+4$	☐	☐	☐

5. The local sponsor of your soccer team needs to know how many soccer balls they might expect to supply you with this year. You calculate that your team will use y≤x+13 soccer balls over the next x months. How many soccer balls might you need this year?

    Ⓐ 20
    Ⓑ 22
    Ⓒ 25
    Ⓓ All of the above

Produce Clear and Coherent Writing

Day 3

6. **What is the purpose of the paragraph below?**

    Bats are the only mammals that fly. They're surprisingly clean animals, spending a majority of their time cleaning their fur.

    Ⓐ Persuade
    Ⓑ Entertain
    Ⓒ Inform
    Ⓓ None of the above

7. **What is the purpose of the sentence below?**

    If you've never visited London, you should plan to go at least once. The city streets are breathtaking, and you'll love British culture.

    Ⓐ To persuade
    Ⓑ To entertain
    Ⓒ To inform
    Ⓓ None of the above

8. **Which of the following sentences is written correctly?**

    Ⓐ We wanted to take the bus to the city but Jeremy called to say that he wood be late.
    Ⓑ We wanted to take the bus to the city, but Jeremy called to say that he would be late
    Ⓒ We wanted to take the bus to the city, but Jeremy called to say that he would be late.
    Ⓓ we wanted to take the bus to the city but Jeremy called to say that he would be late.

9. **True or false?**
    Narratives often use an informal style of writing.

    Ⓐ True
    Ⓑ False

10. Which of the following should be capitalized in a piece of writing?

Ⓐ The beginning letter of a sentence
Ⓑ Proper nouns
Ⓒ Titles
Ⓓ All of the above

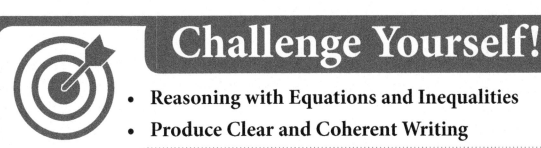

Challenge Yourself!

- **Reasoning with Equations and Inequalities**
- **Produce Clear and Coherent Writing**

https://www.lumoslearning.com/a/dc9-33

Day 3

See Page 7 for Signup details

Building Functions

1. If $f(x) = x^2$ and $g(x) = 2x$, what is $f(g(x))$?

2. What is the equation for your position in terms of t if you are traveling at a constant speed of v and your initial position is x_i?

   Ⓐ $x(t) = x_i - vt$
   Ⓑ $x(t) = x_i vt$
   Ⓒ $x(t) = x_i + vt$
   Ⓓ Not enough information

3. Fill in the table with the 30th number in the sequence.

Term #	Terms in sequence
1	1
2	6
3	11
4	16
30	

4. Your allowance increases by $2.50 every year and at the end of the first year your earning is $10. Which could represent your allowance as a function of years.

   Ⓐ $F(n) = 10.00 + 2.50n$
   Ⓑ $F(n) = 7.50 + 2.50n$
   Ⓒ $F(n) = 7.50n + 2.50$
   Ⓓ $F(n) = 10.00n + 2.50$

5. How is the graph of $f(x) = x + 7$ different from $g(x) = x + 12$?

   Ⓐ When f(x) is shifted up 5 units, g(x) will be obtained
   Ⓑ g(x) is obtained by shifting f(x) down 5 units
   Ⓒ When g(x) is shifted up 5 units, f(x) will be obtained
   Ⓓ f(x) is obtained by shifting g(x) up 5 units

Day 4

6. Which of the following is a simple sentence?

Ⓐ The dog ran away and found his previous owners.
Ⓑ The dog ran away, but we found him.
Ⓒ The dog ran away because I forgot to close the gate.
Ⓓ Because I forgot to close the gate, the dog ran away, but we found him.

7. Which of the following is a compound sentence?

Ⓐ Sam hates doing homework even though her parents push her to do her best.
Ⓑ Sam hates doing homework, but her parents push her to do her best.
Ⓒ Sam hates doing homework and disappointing her parents.
Ⓓ Sam hates doing homework until she goes to sleep, but her parents push her to do her best.

8. Which sentence uses an introductory phrase correctly?

Ⓐ After the game ended, we went to the pizza restaurant.
Ⓑ After the game, we went to the pizza restaurant.
Ⓒ After the game ended we went to the pizza restaurant.
Ⓓ After the game we went to the pizza restaurant.

9. Which of the following is NOT a part of the editing stage of the writing process?

Ⓐ Looking over grammar
Ⓑ Checking capitalization
Ⓒ Replacing weak verbs with stronger verbs
Ⓓ Proofreading punctuation

10. Which of the following is NOT a part of the revision stage of the writing process?

Ⓐ Reordering paragraphs
Ⓑ Revising sentences for variety
Ⓒ Correcting spelling mistakes
Ⓓ Adding information

 **Challenge Yourself!**

- **Building Functions**
- **Develop and Strengthen Writing**

  https://www.lumoslearning.com/a/dc9-34

Day 4

See Page 7 for Signup details

Day 5

1. **What is the inverse function of $f(x)=\sqrt{3x}-2$?**

   Ⓐ $f^{-1}(x)=\dfrac{(x-2)^2}{-3}$

   Ⓑ $f^{-1}(x)=\dfrac{(x-2)^2}{3}$

   Ⓒ $f^{-1}(x)=\dfrac{(x+2)^2}{3}$

   Ⓓ $f^{-1}(x)=\dfrac{(x+2)^2}{-3}$

2. **What is the inverse of the function $f(x)=3^x$?**

   Ⓐ $f^{-1}(x)=\log_3 x$
   Ⓑ $f^{-1}(x)=\log x$
   Ⓒ $f^{-1}(x)=x^2$
   Ⓓ Inverse does not exist

3. **If triangle DEF is a dilation of triangle ABC, what is the scale factor?**

   Ⓐ 0.5
   Ⓑ 1
   Ⓒ 2
   Ⓓ -2

   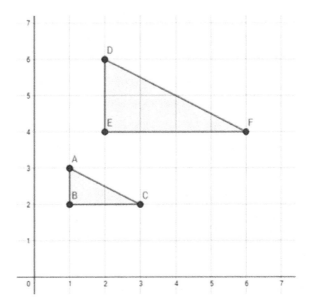

4. **The ratio of length in an image to the preimage is known as what?**

   Ⓐ Dilation
   Ⓑ Scale factor
   Ⓒ Reflection
   Ⓓ Orientation

5. **A line segment has the endpoints (3, 2) and (1, 2). Which pair of endpoints could be for an image that was dilated by a scale factor of 2 with an unknown center?**

Ⓐ (0, 1) and (2, 3)
Ⓑ (2, 2) and (0, 2)
Ⓒ (0, 2) and (4, 2)
Ⓓ (1, 2) and (3, 2)

Develop and strengthen writing (Contd.)

Day 5

6. **Arrange the terms following the correct order of the writing process.**

Ⓐ Drafting
Ⓑ Revising
Ⓒ Proofreading
Ⓓ Planning
Ⓔ Publishing

7. **Why is it important to know the writing process to create a polished piece of work? Explain your answer.**

8. **Which state of the writing process tends to be the most complex and require the most amount of time?**

9. **If you are in the revising stage, you should focus on punctuation and capitalization.**

   Ⓐ True
   Ⓑ False

10. **Which of the following sentences shows evidence of having been proofread correctly:**

    Ⓐ When we think about the minimum age for voting, the first question that comes to our mind is: How young is too young.
    Ⓑ When we think about the minimum age for voting, the first question that comes to our mind is; How young is too young!
    Ⓒ When we think about the minimum age for voting, the first question that comes to our mind is: How young is too young?
    Ⓓ When we think about the minimum age for voting. The first question that comes to our mind is: How young is too young?

Challenge Yourself!

- Building Functions
- Develop and Strengthen Writing(Contd.)

https://www.lumoslearning.com/a/dc9-35

Day 5

See Page 7 for Signup details

Do The Subjects You Choose Really Matter?
How to pick the right high school classes for your dream career

While it may seem a long way away, your grades and high school transcript are crucial parts of the college admissions process, and therefore a step towards your dream career. However, it is important not to take a high school class just because you think it will look good on a college application. Here are some factors to consider when choosing high school classes:

- **Create a balanced set of classes**

While it may be tempting to overload your schedule to look good or take only easy classes to ace them all, it is important to find a balance. Generally, it is a good idea to take classes in English, science, math, the social sciences, and a foreign language. You should be choosing honors classes when you have the capability to do well, however overdoing it and only getting Cs won't do you any favors, either. Talk to your school guidance counselor and even your parents to help get some level of advice on what a good class-load looks like for you.

- **Don't take classes for the wrong reasons**

Taking a class just because your friends are is not a good reason, just as taking an AP class just to look advanced is also not a good reason. You should focus on your strengths, interests and creating a realistic class schedule that you can thrive with. It goes to say, that if you hate it you probably won't apply yourself very hard.

- **Consider college admission prerequisites**

If you have looked into your dream college courses, you surely will have realized there are certain requirements for each school or course. Certain colleges may require you to take a foreign language, for example, or a certain level of math. By having a general idea of schools you'd like to attend, you can design a high school transcript that will be advantageous come college admissions time.

- **Pursue your real interests**

This one seems simple, but you are more likely to succeed at classes you thoroughly enjoy. You may think you want to study medicine in the future, but if you really hate your chemistry class in high school it may be a sign to consider a different path. The same goes for the opposite! If there is a class that you really love, research careers that stem from this knowledge. It is likely that if you like it as a high school class, you will find a fulfilling career in the future.

Week 7 - PSAT/NMSQT Prep

- Math
- Evidence Based Writing & Language

https://www.lumoslearning.com/a/slh9-10

See Page 7 for Signup details

Weekly Fun Summer Photo Contest

Take a picture of your summer fun activity and share it on Twitter or Instagram

Use the **#SummerLearning** mention

@LumosLearning on Twitter or

@lumos.learning on Instagram

Tag friends and increase your chances of winning the contest

Participate and stand a chance to **WIN $50 Amazon gift card!**

1. Which of the following image A'B'C' is similar to triangle ABC?

Ⓐ

Ⓑ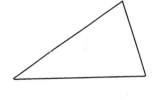

Ⓒ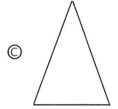

Ⓓ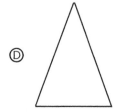

2. Which of the following triangles would not be similar to triangle ABC?

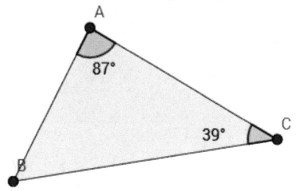

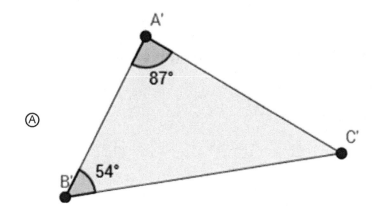

Ⓐ

Ⓒ

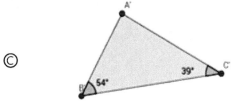

Ⓑ

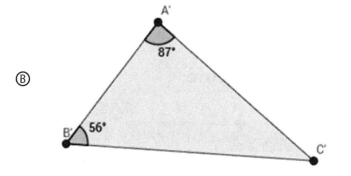

Ⓓ

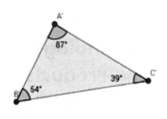

3. What does the AA stand for in the AA similarity theorem?

Ⓐ Acute Angles
Ⓑ Adjacent Angles
Ⓒ Angle Angle
Ⓓ Adjacent Acute

4. Given that $\overline{AB}||\overline{ST}$, are triangles ABX and STX similar?

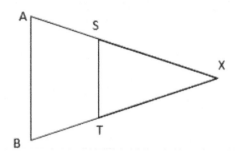

Ⓐ No, they are not similar
Ⓑ No, they are congruent
Ⓒ Yes, they are similar
Ⓓ Not enough information given

5. In the figure below, if the length of side XZ is 9, what is the length of side YZ?

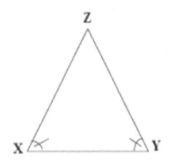

Ⓐ 7
Ⓑ 9
Ⓒ 18
Ⓓ 4.5

Use Technology to Produce, Publish, and Update Writing Products

Day 1

6. Which of the following is a correct parenthetical citation in MLA?

Ⓐ Later on, he reflected on his past by saying, "Turning to the Dark Side seemed like the only way I could stop the pain I knew was coming" (There and Back Again 83).
Ⓑ Later on, he reflected on his past by saying, "Turning to the Dark Side seemed like the only way I could stop the pain I knew was coming" (Skywalker 83).
Ⓒ Later on, he reflected on his past by saying, "Turning to the Dark Side seemed like the only way I could stop the pain I knew was coming." (Skywalker 83)
Ⓓ Later on, he reflected on his past by saying, "Turning to the Dark Side seemed like the only way I could stop the pain I knew was coming." (Skywalker 83).

7. **MLA stands for**

Ⓐ Modern Language Association
Ⓑ Modern Language Arts
Ⓒ Main Linguistic Association
Ⓓ Multiple Language Assets

8. **Which of the following is an appropriate software to type your writing?**

Ⓐ Google Documents
Ⓑ Microsoft Word
Ⓒ Apple Pages
Ⓓ All of the above

9. **Taking information from a source and putting it into your own words is known as**

Ⓐ quoting
Ⓑ summarizing
Ⓒ paraphrasing
Ⓓ phrasing

10. **Information quoted from a source must be cited.**

True or False? Write your answer in the box below.

Challenge Yourself!

- Similarity, Right Triangles, & Trigonometry
- Use Technology to Produce, Publish, and Update Writing Products

https://www.lumoslearning.com/a/dc9-36

Day 1

See Page 7 for Signup details

1. The congruency markings on one of the triangles in the figure below are incorrect. Two of the three triangles in the figure are congruent. Which congruency statement correctly says which two triangles are congruent?

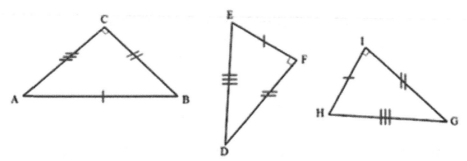

Ⓐ ΔABC≅ΔHGI
Ⓑ ΔDEF≅ΔABC
Ⓒ ΔDEF≅ΔGHI
Ⓓ ΔDEF≅ΔIGH

2. The congruency markings on one of the triangles in the figure below are incorrect. Two of the three triangles in the figure are congruent. Which congruency statement correctly says which two triangles are congruent?

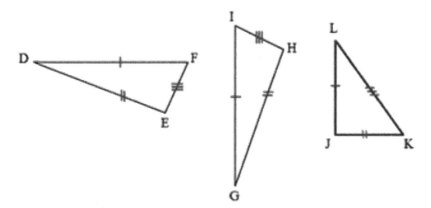

Ⓐ ΔJKL≅ΔHGI
Ⓑ ΔDEF≅ΔGHI
Ⓒ ΔDEF≅ΔJKL
Ⓓ ΔDEF≅ΔIGH

3. What is the triangle similarity statement that can be used to discuss the similarity of the two triangles below?

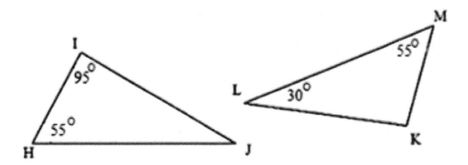

- Ⓐ ΔHIJ~ΔLKM
- Ⓑ ΔHIJ~ΔMKL
- Ⓒ The triangles are not similar
- Ⓓ ΔIHJ~ΔLKM

4. In the figure below, ΔABC~ΔXZY. What is m$\overline{YZ}$?

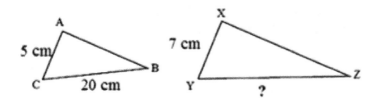

- Ⓐ 14 cm
- Ⓑ 28 cm
- Ⓒ 6 cm
- Ⓓ 28 m

5. In right ΔPQR in the figure below, which trigonometric ratio is equivalent to sin R?

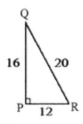

- Ⓐ cos Q
- Ⓑ tan R
- Ⓒ tan P
- Ⓓ sin Q

Conduct Short and Sustained Research Project

6. Finding sources for a research project should be guided by a clear

    Ⓐ thesis statement
    Ⓑ topic sentence
    Ⓒ controlling idea
    Ⓓ inquiry question

7. Which of the following should you include in a citation of a source on your works cited page?

    Ⓐ Author's name
    Ⓑ Title of source
    Ⓒ Year of publication
    Ⓓ All of the above

8. Which of the following types of writing involves doing research?

    Ⓐ A documentary
    Ⓑ A scientific article
    Ⓒ A biography
    Ⓓ All of the above

9. The following inquiry question would be considered too broad:

    How does recycling impact the environment?

    True or false? Write your answer in the box below.

10. How can you determine if a source is credible?

Linear, Quadratic, & Exponential Models

1. Suppose a cable is connected to a pole 12 meters above the ground and extended away from the base of the pole. The distance from the base of the pole to the end of the cable is 35 meters. What is the angle made by the cable and the ground?

    Ⓐ 20.1°
    Ⓑ 69.9°
    Ⓒ 18.9°
    Ⓓ 71.1°

2. What is the length of the hypotenuse of a right triangle if the legs of the sides are 12 and 35?

    Ⓐ 32.9
    Ⓑ 37
    Ⓒ 50
    Ⓓ 20.5

3. A football coach believes that introducing and increasing weights slowly during strength training is best for his athletes. He starts his athletes bench pressing 81 pounds then increases to 83, 89, 107, and so on each week. This implies what kind of growth?

    Ⓐ Linear
    Ⓑ Exponential
    Ⓒ Geometric
    Ⓓ None of these

4. While traveling home for Christmas Jamie plans to get on the interstate and increase her speed from 0 to 75 in the first minute and a half and then set her cruise control to remain at a constant speed for the first leg of the trip. She doesn't expect any traffic to slow her down. Jamie's speed after she accelerates and before she slows down could best be represented by which of the following?

    Ⓐ Quadratic Function
    Ⓑ Linear Function
    Ⓒ Exponential Function
    Ⓓ None of these

5. Shelly decided to start a home based cookie baking business. During the first year she baked 20 dozen cookies, and the next year she baked 30 dozen. The year after that was 40 dozen. What function could model this success (assuming that x=0 represents the first year)?

Ⓐ $f(x)=20x$
Ⓑ $f(x)=10x+20$
Ⓒ $f(x)=10x$
Ⓓ None of these

Source-Based Writing

Day 3

6. Which of the following is considered a primary source?

Ⓐ A biography
Ⓑ An autobiography
Ⓒ An encyclopedia
Ⓓ A documentary

7. Which of the following is considered a secondary source?

Ⓐ An autobiography
Ⓑ A personal letter
Ⓒ An interview
Ⓓ A biography

8. What is the difference between a primary source and a secondary source?

9. **What is the primary purpose of an in-text citation?**

Ⓐ To tell the reader where the information came from
Ⓑ To remind the author where the information came from
Ⓒ To make an author's text look more aesthetic
Ⓓ To provide the reader with a link to find more information

10. **Why is it important to gather relevant information from multiple sources?**

Challenge Yourself!

- **Similarity, Right Triangles, & Trigonometry**
- **Source-Based Writing**

https://www.lumoslearning.com/a/dc9-38

Day 3

See Page 7 for Signup details

LumosLearning.com

Day 4

1. Which of these is a linear function?

   Ⓐ $y=ab^x$
   Ⓑ $y=mx+b$
   Ⓒ $y=a^x$
   Ⓓ $y=ax^2+bx+c$

2. Suppose there are 950 widgets in a warehouse when a new business opened. The business expects to sell 45 of the widgets per week. Which function $I(w)$ represents the number of widgets that will be in the warehouse after selling them for w weeks?

   Ⓐ $I(w)=-950+45w$
   Ⓑ $I(w)=950-45w$
   Ⓒ $I(w)=950+45w$
   Ⓓ $I(w)=45-950w$

3. Suppose the Thespian troop at a high school spent $855 to produce a play. If the tickets cost $20 each, which function $P(t)$ represents the net profit of the play, with t representing the number of tickets sold?

   Ⓐ $P(t)=-20t+855$
   Ⓑ $P(t)=20t-855$
   Ⓒ $P(t)=20t+855$
   Ⓓ $P(t)=855t-20$

4. What is the length, in inches, of an arc subtended by an angle of 36° on a circle with a radius of 15 inches?

   Ⓐ 540
   Ⓑ 36π
   Ⓒ 36
   Ⓓ 3π

5. What is the length, in centimeters, of an arc subtended by an angle of $\frac{2\pi}{7}$ radians on a circle with a radius of 14 centimeters?

   Ⓐ 14
   Ⓑ π
   Ⓒ 4π
   Ⓓ 7π

Citing Text-Evidence

Smartphones

(1) With the current prevalence of smartphones, owned by young and old alike, it might come as a surprise that smartphone technology is a fairly recent development. Steve Jobs, the former CEO and visionary of Apple who died in October 2011 of cancer revealed the company's first iPhone on January 9, 2007. Combining the ability to make a phone call, text messaging, internet browsing, music streaming, and a number of other capabilities, the iPhone and other smartphones that followed in the iPhone's footsteps quickly became commonplace as the go-to informational access device for people all over the world. Though smartphones offer a significant number of conveniences, the overreliance on the technology has become a growing concern among teenagers and adults.

(2) In classrooms all over the country, teachers are fighting a battle for the attention of the students in their classes. The opposition? The smartphone with almost unlimited options resting in the students' pockets, or, more likely, in the students' hands. With access to videos, games, and messaging services that allow them to connect with their friends, even if the friend is in the same classroom, students often find their smartphone screen to be more engaging than the lesson on gerunds the teacher has to offer. And though teachers might try to solve the problem by taking up a student's cell phone, a surprising amount of anxiety is felt by the student when he or she is disconnected from the device that makes them feel most connected to the world at large. "Taking it away doesn't solve the problem," said 9th-grade history teacher, Morgan Giles. She continued to say that taking the students' smartphones just created a power struggle. "There has to be a better way," she said.

(3) Studies have also shown that students are spending so much time on their screens that it's affecting their ability to interact with the real world around them. Students are losing sleep from staying up late to play games on their smartphone, according to a study done by the American Academy of Sleep Medicine. Even family connections are being negatively impacted by smartphone usage, as families gather around the dinner table, faces buried in their smartphones rather than engaging in conversation with the people around them.

(4) Some might argue that smartphones increase student safety. Having a smartphone, after all, keeps students connected to their parents, and they always have a way to contact authorities at the touch of a button. While these are fair arguments, the benefits don't erase the negative effects teenagers are suffering from the overuse of their smartphones.

(5) While smartphones have certainly made many things more convenient, parents would do well to teach their children how to better handle something so small that asks for so much of their attention.

6. Which of the following best describes the tone of the passage?

Ⓐ Lighthearted
Ⓑ Cautionary
Ⓒ Dire
Ⓓ Melancholy

7. Briefly summarize the passage.

8. What argument does the author make?

Ⓐ Smartphones provide a number of conveniences to their users.
Ⓑ Smartphones are dangerous and should be eliminated.
Ⓒ Smartphones are designed to be a distraction to their users.
Ⓓ Despite their benefits, smartphones are proving to be a distraction to teenagers.

9. Which paragraph includes the counterclaim?

Ⓐ Paragraph 1
Ⓑ Paragraph 2
Ⓒ Paragraph 3
Ⓓ Paragraph 4

10. What kind of evidence does the author use to support his argument?

Challenge Yourself!

- **Linear, Quadratic, & Exponential Models**
- **Citing Text-Evidence**

https://www.lumoslearning.com/a/dc9-39

Day 4

See Page 7 for Signup details

Day 5

1. In the unit circle, one can see that $\sin(\frac{\pi}{3}) = \frac{\sqrt{3}}{2}$. What is the value of $\cos(\frac{\pi}{3})$?

    Ⓐ $\frac{\sqrt{3}}{2}$

    Ⓑ $\frac{1}{2}$

    Ⓒ $\frac{2\sqrt{3}}{3}$

    Ⓓ 2

2. In the unit circle, one can see that $\cos(\frac{\pi}{3}) = \frac{1}{2}$. What is the value of $\sin(\frac{\pi}{3})$?

    Ⓐ $-\frac{\sqrt{3}}{2}$

    Ⓑ $\frac{1}{2}$

    Ⓒ $\frac{\sqrt{3}}{2}$

    Ⓓ $-\frac{1}{2}$

3. Using special triangles to find the value of trigonometric functions, what is $\cos\frac{\pi}{6}$?

    Ⓐ 2

    Ⓑ $\frac{1}{2}$

    Ⓒ $\frac{\sqrt{3}}{2}$

    Ⓓ $\frac{1}{3}$

4. Using special triangles to find the value of trigonometric functions, what is $\sin 135°$?

    Ⓐ $-\frac{\sqrt{2}}{2}$

    Ⓑ $\frac{\sqrt{2}}{2}$

    Ⓒ $-\frac{1}{2}$

    Ⓓ $\frac{1}{2}$

5. Using the unit circle, find the value of $\sin\frac{\pi}{3}$ and $\cos\frac{\pi}{3}$. What is the value of $\sin(\frac{\pi}{2}+\frac{\pi}{3})$ equivalent to?

Ⓐ $\sin\frac{\pi}{3}$

Ⓑ $\tan\frac{\pi}{3}$

Ⓒ $\cos\frac{\pi}{3}$

Ⓓ $\cos\frac{5\pi}{3}$

Day 5

Read the passage and answer the questions (Q6 & Q7)

Teenagers often board school buses before most people have even gotten out of bed, causing chronic sleep deprivation, a major public health concern. This should be scary enough to debate national reform to school schedules. Now researchers state that starting school later in the morning, no earlier than 8:30 am, would actually have financial benefits as well. Students who are better rested perform better in school; this translates to better career opportunities later in life which has a positive effect on the national economy. Also, fewer car crashes would result from drowsy teenagers driving to school before they are fully awake.

6. **What type of writing is the above passage?**

Ⓐ narrative
Ⓑ informational
Ⓒ argument
Ⓓ Could be both B and C.

7. **Which is the central idea of the above passage?**

Ⓐ School buses pick up students too early in the morning.
Ⓑ Chronic sleep deprivation is a major public health concern.
Ⓒ Better rested students would result in better academic performance.
Ⓓ Implementing later school start times would have a positive health and economic effect.

Merry Autumn- by Paul Laurence Dunbar

Now purple tints are all around;
The sky is blue and mellow;
And e'en the grasses turn the ground
From modest green to yellow...

A butterfly goes winging by;
A singing bird comes after;
And Nature, all from earth to sky,
Is bubbling o'er with laughter...

The earth is just so full of fun
It really can't contain it;
And streams of mirth so freely run
The heavens seem to rain it...

Why, it's the climax of the year,—
The highest time of living!—
Till naturally its bursting cheer
Just melts into thanksgiving.

Read the excerpt from the poem "Merry Autumn" by Paul Laurence Dunbar, then answer the questions(Q8 & Q9).

8. **The phrases "purple tints", "modest green", "singing bird", "bubbling o'er with laughter" are examples of...**

    Ⓐ personification.
    Ⓑ sensory details.
    Ⓒ metaphors.
    Ⓓ onomatopoeia.

9. **Which line from the poem best captures the author's main idea?**

    Ⓐ "Now purple tints are all around"
    Ⓑ "And Nature, all from earth to sky/Is bubbling o'er with laughter"
    Ⓒ "Why, it's the climax of the year/The highest time of living"
    Ⓓ "Till naturally its busting cheer/Just melts into thanksgiving"

Read the excerpt from My Antonia by Willa Cather, then answer the question.

"I had a sense of coming home to myself, and of having found out what a little circle man's experience is. For Antonia and for me, this had been the road of Destiny; had taken us to those early accidents of fortune which predetermined for us all that we can ever be. Now I understood that the same road was to bring us together again. Whatever we had missed, we possessed together the precious, the incommunicable past."

10. What is the meaning of the word "predetermined" as it is used in this context?

Ⓐ an accident or unfortunate circumstance
Ⓑ to be decided in advance
Ⓒ easy to agree upon
Ⓓ something that is flexible or unsettled

Challenge Yourself!

- **Trigonometric Functions**
- **Citing Text-Evidence(Contd.)**

https://www.lumoslearning.com/a/dc9-40

Day 5

See Page 7 for Signup details

How to Find Scholarships for College

College might feel like a long way off, especially if you just started high school, but it's never too early to begin thinking about majors you might be interested in and the schools that offer those majors.

You'll also want to begin thinking about how you're going to pay for college. There are different options available and you will want to talk with your parents to decide what works best for you and your family.

Most colleges offer some type of financial aid package. These usually include loans, which is money that you'll have to pay back, and grants and scholarships, if you qualify for these. Grants are given to students based on financial need and never have to be paid back. Scholarships are awarded based on merit (grades) or talent (basketball, dance, etc.) and just like grants, never have to be paid back.

This is why scholarships are so competitive. It's free money and everyone wants it. There are thousands of scholarships out there and you should try to get as many as you can to help cover the cost of tuition, your dorm, food, and other expenses like books and fees.

So where do you find scholarships?

1. Talk to your guidance counselor at your high school. They have the most up-to-date information on local scholarships as well as websites you can search online.
2. Check where your parents work. Sometimes companies will have scholarships for their employees' dependent children.
3. Does your church or religious organization offer scholarships for its members?
4. The US Department of Labor scholarship search tool is a good resource that allows you to search for a wide variety of scholarships. Click on https://www.careeronestop.org/Toolkit/Training/find-scholarships.aspx to be redirected to their website.
5. Explore your local library. Most libraries have a reference section that might include books on scholarships. You can also talk with the librarian and they will direct you to the books that they have available for their patrons on this subject.

To help you win scholarships make sure to apply to as many scholarships as possible and in different monetary amounts. The high dollar amounts tend to be the most competitive. Also, keep in mind deadlines.

For most scholarship applications you'll need your high school transcripts, both unofficial and official, and your SAT or ACT scores. Some will also want letters of recommendation from teachers, coaches, or someone you've worked or volunteered with, and sometimes you'll need to write an essay. The most important thing is to follow the scholarship application directions. It's important to follow the directions on each scholarship application you fill out and submit everything that they request. If they want a 500-word essay, write a 500-word essay. Going over the word count just because you want to elaborate on a topic could disqualify you.

The sooner you start thinking about the cost of school the more time you'll have to save and find ways to pay for it. Going to college is an investment in your future, and for many students, it's the next step to gain the necessary skills and education needed to pursue their career field.

Week 8 - PSAT/NMSQT Prep

- **Math**
- **Evidence Based Writing & Language**

https://www.lumoslearning.com/a/slh9-10

See Page 7 for Signup details

Weekly Fun Summer Photo Contest

Take a picture of your summer fun activity and share it on Twitter or Instagram

Use the **#SummerLearning** mention

@LumosLearning on Twitter or

@lumos.learning on Instagram

Tag friends and increase your chances of winning the contest

Participate and stand a chance to WIN $50 Amazon gift card!

Symmetry and Odd-Even Relationships in the Unit Circle

1. The value of $\sin\frac{\pi}{6}$ is $\frac{1}{2}$. Using the symmetry and odd-even relationships in the unit circle, what is the value of $\sin\frac{5\pi}{6}$ equivalent to?

   Ⓐ $\cos\frac{11\pi}{6}$

   Ⓑ $\frac{\sqrt{3}}{4}$

   Ⓒ $-\frac{\sqrt{3}}{2}$

   Ⓓ $\sin\frac{\pi}{6}$

2. The value of $\cos\frac{11\pi}{6}$ is $\frac{\sqrt{3}}{2}$. Using the symmetry and odd-even relationships in the unit circle, what is the value of $\cos\frac{\pi}{6}$?

   Ⓐ $-\frac{\sqrt{3}}{2}$

   Ⓑ $\frac{1}{2}$

   Ⓒ $\frac{\sqrt{3}}{2}$

   Ⓓ $-\frac{1}{2}$

3. The value of $\tan\frac{\pi}{3}$ is $\sqrt{3}$. Using the symmetry and odd-even relationships in the unit circle, what is the value of $\tan(-\frac{\pi}{3})$?

   Ⓐ $-\sqrt{3}$
   Ⓑ $\sqrt{3}$
   Ⓒ -2
   Ⓓ 2

4. The value of $\sin 60°$ is $\frac{\sqrt{3}}{2}$. Using the symmetry and odd-even relationships in the unit circle, what is the value of $\sin 120°$?

  Ⓐ $\frac{\sqrt{3}}{2}$

  Ⓑ $\frac{\sqrt{2}}{3}$

  Ⓒ $-\frac{\sqrt{3}}{2}$

  Ⓓ $\frac{1}{2}$

5. The value of $\cos 45°$ is $\frac{\sqrt{2}}{2}$. Using the symmetry and odd-even relationships in the unit circle, what is the value of $\cos 135°$?

  Ⓐ $\frac{\sqrt{2}}{2}$

  Ⓑ $\frac{1}{2}$

  Ⓒ $-\frac{\sqrt{2}}{2}$

  Ⓓ $-\frac{1}{2}$

Grammar and Usage

Day 1

6. Which phrase represents an adverb phrase in the following sentence: "He placed the present by the birthday cake."

  Ⓐ He placed
  Ⓑ the present
  Ⓒ by the birthday cake
  Ⓓ placed the present

7. Which phrase represents an adverb clause in the following sentence: "You can start the oven while I finish making the dough."

  Ⓐ You can start
  Ⓑ start the oven
  Ⓒ making the dough
  Ⓓ while I finish making the dough

8. What word is described by the adverb extremely in the following sentence:
    "The present he purchased for her birthday was extremely expensive."

    Ⓐ present
    Ⓑ purchased
    Ⓒ birthday
    Ⓓ expensive

9. Which type of clause is represented in the following sentence?
    "Humans and insects are similar in that they both need to breathe to survive, and <u>both breathe out carbon dioxide</u>."

    Ⓐ independent clause
    Ⓑ dependent clause
    Ⓒ noun clause
    Ⓓ relative clause

10. Which type of clause is represented in the following sentence?
    "Jane's English teacher, <u>who is also in charge of the school yearbook</u>, is retiring next year."

    Ⓐ independent clause
    Ⓑ adverbial clause
    Ⓒ noun clause
    Ⓓ relative clause

Amplitude, Frequency, and Midlines

1. What is the amplitude of the function $y = 9 \sin(x + \frac{\pi}{2})$?

    Ⓐ $\frac{9}{2}$

    Ⓑ 18

    Ⓒ 9

    Ⓓ -9

2. What is the equation of the mid-line of the trigonometric function $y = \frac{4}{5} \sin(2x - \pi) - 5$?

    Ⓐ $y = 5$

    Ⓑ $y = -5$

    Ⓒ $y = \frac{4}{5}$

    Ⓓ $y = 0$

3. Which of the following trigonometric functions has a mid-line with an equation of $y = \frac{7}{2}$?

    Ⓐ $y = 9\cos(x - \pi) - \frac{7}{2}$

    Ⓑ $y = -\frac{2}{7}\cos(x - \pi) + \frac{7}{2}$

    Ⓒ $y = \frac{7}{2}\sin(x + \pi) - \frac{\pi}{2}$

    Ⓓ $y = -\frac{7}{2}\cos(5x) + \frac{5\pi}{3}$

4. Which one of the following functions has a period of $\frac{8\pi}{3}$?

    Ⓐ $y = 2\sin(\frac{4x}{3}) + 5$

    Ⓑ $y = 4\tan(\frac{3x}{4}) - 1$

    Ⓒ $y = 2\sin(\frac{3x}{4}) - 5$

    Ⓓ $y = \frac{8}{3}\sin(x) - 2$

5. What is the amplitude of the function $y = -2\sin(x-\pi)+3$?

Ⓐ 2
Ⓑ 4
Ⓒ -2
Ⓓ π

Day 2

Determining Unknown Words

6. Read the sentence. Based on the context, determine the meaning of the underlined word. "It was the Student Council president's job to <u>disseminate</u> information about the school events to the students in her class, making sure everyone was informed."

Ⓐ spread
Ⓑ keep secret
Ⓒ disorganize
Ⓓ decide

7. Read the sentence. Based on the context, determine the meaning of the underlined word. "In countries where the government is controlled by a dictator, officials usually jail <u>dissidents</u> who disagree with the laws."

Ⓐ criminals
Ⓑ protesters
Ⓒ teachers
Ⓓ government officials

8. Read the sentence. Based on the context, determine the meaning of the underlined word. "The teacher kept the class running smoothly by <u>facilitating</u> the lesson sequence, keeping everyone on task."

Ⓐ controlling
Ⓑ reporting
Ⓒ organizing
Ⓓ canceling

9. Repeat, rename, and rebuild all begin with the prefix re-. What is the meaning of the prefix re-?

   Ⓐ not
   Ⓑ before
   Ⓒ again
   Ⓓ under

10. Unfinished, unskilled, and unfriendly all begin with the prefix un-. What is the meaning of the prefix un-?

   Ⓐ not
   Ⓑ before
   Ⓒ again
   Ⓓ under

Challenge Yourself!

- Amplitude, Frequency, and Midlines
- Determining Unknown Words

https://www.lumoslearning.com/a/dc9-42

Day 2

See Page 7 for Signup details

Day 3

1. The graph of the cosine function is shown below.

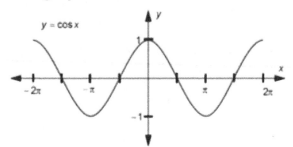

Which of the following domain intervals will allow the inverse of the cosine to be a function?

Ⓐ $[-2\pi, 0]$
Ⓑ $[0, \pi]$
Ⓒ $[0, 2\pi]$
Ⓓ $[-\pi, \pi]$

2. The graph of the sine function is shown below.

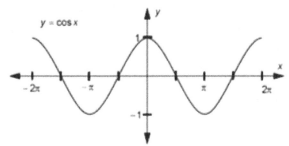

Which of the following domain intervals will allow the inverse of the sine to be a function?

Ⓐ $[-\pi, \pi]$
Ⓑ $[0, 2\pi]$
Ⓒ $[-\dfrac{\pi}{2}, \dfrac{\pi}{2}]$
Ⓓ $[\pi, -\pi]$

3. The graph of the tangent function is shown below.

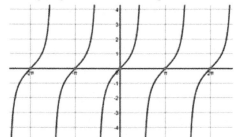

Which of the following domain intervals will allow the inverse of the tangent to be a function?

Ⓐ $[-\pi,\pi]$
Ⓑ $[0,2\pi]$
Ⓒ $[-\dfrac{\pi}{2},\dfrac{\pi}{2}]$
Ⓓ $[\pi,-\pi]$

4. The graph of the cotangent function is shown below.

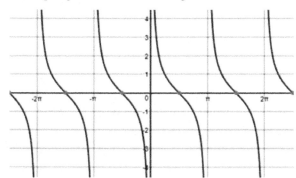

Which of the following domain intervals will allow the inverse of the cotangent to be a function?

Ⓐ $[-2\pi,0]$
Ⓑ $[0,\pi]$
Ⓒ $[0,2\pi]$
Ⓓ $[-\pi,\pi]$

5. The graph of the cosecant function is shown below.

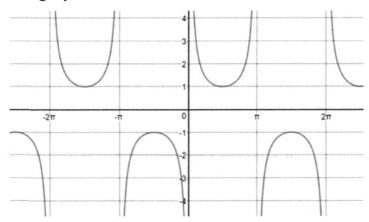

Which of the following domain intervals will allow the inverse of the cosecant to be a function?

Ⓐ $[-2\pi,0]$

Ⓑ $[0,\pi] - \{\frac{\pi}{2}\}$

Ⓒ $[-\frac{\pi}{2},\frac{\pi}{2}] - \{0\}$

Ⓓ $[-\pi,\pi] - \{\frac{\pi}{2}\}$

Day 3

Academic and Domain-Specific Vocabulary

6. Choose the academic term that best completes the sentence.
"China's _____ is largely dependent on manufacturing with 80% of their exports being manufactured goods."

Ⓐ government
Ⓑ research
Ⓒ bank
Ⓓ economy

7. Choose the academic term that best completes the sentence.

"The main task of the ____ branch of the United States government is to make laws."

Ⓐ executive
Ⓑ legislative
Ⓒ congressional
Ⓓ judicial

8. Choose the academic term that best completes the sentence.

    "Clare conducted a/an ____ of the way children learn to read as the subject of her Master's thesis to complete her second college degree."

    Ⓐ analysis
    Ⓑ research
    Ⓒ data
    Ⓓ summary

9. Choose the academic term that best completes the sentence.

    "For teenagers, whose brains are not yet fully developed, the ____ of rational and irrational thinking is not always clear."

    Ⓐ principles
    Ⓑ concepts
    Ⓒ roles
    Ⓓ factors

10 Choose the academic term that best completes the sentence.

    "The police had to free the suspected burglar because there was not enough ___ to convict him."

    Ⓐ legal
    Ⓑ theory
    Ⓒ evidence
    Ⓓ witness

Challenge Yourself!

- Inverses of the Trigonometric Functions
- Academic and Domain-Specific Vocabulary

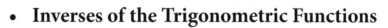

https://www.lumoslearning.com/a/dc9-43

Day 3

See Page 7 for Signup details

Solving Trigonometric Equations Using Inverses

1. Solve the trigonometric equation below.

   $1 - \cos\theta = \dfrac{1}{2}$

   What is the solution if $\pi \leq \theta \leq 2\pi$?

   Ⓐ $\dfrac{\pi}{3}$

   Ⓑ $\dfrac{4\pi}{3}$

   Ⓒ $\dfrac{5\pi}{3}$

   Ⓓ $\dfrac{2\pi}{3}$

2. Solve the trigonometric equation below.
   $4\cos^2\theta = 1$

   What is the solution if $\pi \leq \theta \leq \dfrac{3\pi}{2}$?

   Ⓐ $\dfrac{\pi}{3}$

   Ⓑ $\dfrac{4\pi}{3}$

   Ⓒ $\dfrac{5\pi}{3}$

   Ⓓ $\dfrac{2\pi}{3}$

3. Solve the trigonometric equation below.
   $\tan(2\theta) = -1$

   What is the solution if $-\dfrac{\pi}{2} \leq \theta \leq \pi$?

   Ⓐ $\dfrac{3\pi}{8}$

   Ⓑ $\dfrac{11\pi}{8}$

   Ⓒ $\dfrac{15\pi}{8}$

   Ⓓ $\dfrac{7\pi}{8}$

4. Solve the trigonometric equation below.
   $\cot \theta = 1$

   What is the solution if $0 \le \theta \le \dfrac{\pi}{2}$**?**

   Ⓐ $\dfrac{3\pi}{4}$

   Ⓑ $\dfrac{\pi}{4}$

   Ⓒ $\dfrac{5\pi}{4}$

   Ⓓ $\dfrac{7\pi}{4}$

5. Solve the trigonometric equation below.
   $2\sin \theta + \sqrt{3} = 0$

   What is the solution if $\dfrac{3\pi}{2} \le \theta \le 2\pi$**?**

   Ⓐ $\dfrac{3\pi}{4}$

   Ⓑ $\dfrac{5\pi}{4}$

   Ⓒ $\dfrac{5\pi}{3}$

   Ⓓ $\dfrac{4\pi}{3}$

Write informative/explanatory texts

Day 4

6. When you are writing to inform or explain you should:

   Ⓐ List a lot of really good facts
   Ⓑ Use a lot of opinions
   Ⓒ Write in first person
   Ⓓ None of the above

7. Which of the following is the MOST important part of informing or explaining?

   Ⓐ Opinions
   Ⓑ Facts
   Ⓒ Hyperbole
   Ⓓ None of the above

8. **Which of the following pieces of writing is MOST likely to be informative?**

   Ⓐ A set of directions to the mall.
   Ⓑ An explanation of how to knit.
   Ⓒ An explanation of the causes of WWI.
   Ⓓ All of the above.

9. **When you organize your writing for an informational text, you should use**

   Ⓐ Big paragraphs
   Ⓑ Bullet point lists
   Ⓒ Complex sentences
   Ⓓ None of the above

10. **True or false, this is a sentence you would see in an informational text:**

    Get down to Marty's Auto Shop and get this amazing deal on tires today!

Trigonometric Functions

1. If θ is in quadrant III, and $\cos\theta = -\dfrac{12}{13}$, what is $\tan\theta$?

   Ⓐ $-\dfrac{12}{5}$

   Ⓑ $\dfrac{12}{5}$

   Ⓒ $\dfrac{12}{13}$

   Ⓓ $-\dfrac{5}{12}$

2. If θ is in quadrant II, and $\sin\theta = -\dfrac{7}{25}$, what is $\cos\theta$?

   Ⓐ $\dfrac{24}{25}$

   Ⓑ $-\dfrac{7}{24}$

   Ⓒ $-\dfrac{24}{25}$

   Ⓓ $\dfrac{24}{7}$

3. If θ is in quadrant I, and $\sin\theta = \dfrac{3}{5}$, what is $\cos\theta$?

   Ⓐ $-\dfrac{4}{5}$

   Ⓑ $\dfrac{5}{3}$

   Ⓒ $\dfrac{5}{4}$

   Ⓓ $\dfrac{4}{5}$

4. Recall that $\cos(\alpha + \beta) = \cos\alpha\cos\beta - \sin\alpha\sin\beta$. What is the exact value of $\cos 75°$?

   Ⓐ $\dfrac{\sqrt{6}}{4} + \dfrac{\sqrt{2}}{4}$

   Ⓑ $\dfrac{\sqrt{6}}{4} - \dfrac{\sqrt{2}}{4}$

   Ⓒ $\dfrac{\sqrt{2}}{4} - \dfrac{\sqrt{6}}{4}$

   Ⓓ $\dfrac{\sqrt{3}}{2} - \dfrac{\sqrt{2}}{2}$

5. Recall that $\tan(\alpha - \beta) = \dfrac{\tan\alpha - \tan\beta}{1 + \tan\alpha\tan\beta}$. What is the exact value of $\tan 15°$?

Ⓐ $\dfrac{3-\sqrt{3}}{3+\sqrt{3}}$

Ⓑ $\dfrac{3+\sqrt{3}}{3-\sqrt{3}}$

Ⓒ $\dfrac{\sqrt{3}-3}{3+\sqrt{3}}$

Ⓓ $\dfrac{\sqrt{2}}{2} - \dfrac{\sqrt{3}}{2}$

Use technology to produce, publish, and update writing products

Day 5

6. If you copy information from the Internet and use it in your writing without citing it as a source, you are:

Ⓐ Creating
Ⓑ Adjusting
Ⓒ Plagiarizing
Ⓓ None of the above

7. What is it called when you let the reader know where your source came from?

Ⓐ Citing Sources
Ⓑ Repeating sources
Ⓒ Doctoring sources
Ⓓ Bibiliographing sources

8. What is paraphrasing and do you need to cite it?

9. What should be the following characteristics of a good research topic?

Ⓐ focused on a few variables
Ⓑ can be researched and proven with evidence
Ⓒ the topic should be substantial
Ⓓ all of the above

10. Which of the following are styles you can use to cite and reference your paper?

Ⓐ MLA
Ⓑ APA
Ⓒ Chicago
Ⓓ All of the above

Challenge Yourself!

- Solving Trigonometric Equations Using Inverses
- Use technology to produce, publish, and update writing products

https://www.lumoslearning.com/a/dc9-45

Day 5

See Page 7 for Signup details

Preparing a Portfolio for College – the Why of it

If you have not yet started building a portfolio for your college applications, there is no time like the present. But if you have no experience with portfolios, you may not know where to start and what to include?

What is a Portfolio?

A portfolio is a compilation of documents that will show your strengths. It will enhance your resume by showing who you are as a person, your skills, qualifications, training, and other experiences. It will give the reader a more in-depth view of your personality and your work ethic.

Why Do I Need a Portfolio?

For starters, you want to stand out. Not everyone has thought to put together a portfolio despite the importance of having one. It will also improve your interview skills. Your portfolio documents will show the reader what your skills and experiences have been; you will need to elaborate on them when asked the pertinent questions. Having a professional portfolio also indicates your level of organization, communications skills, and any experience you have that will help you to move forward.

What Goes in My Portfolio?

You should save everything you create, and when you are ready to assemble your portfolio, you can start to choose your best work. You should have items that will reflect your skills and experiences.

1. Statement of Purpose: You should introduce your portfolio with a statement explaining your portfolio's purpose and contents. It should indicate that it is all your original work and not to be copied for any purpose without permission.

2. Philosophy: A statement about your beliefs regarding college and what you hope to achieve.

3. Goals: Include a statement of your goals for the next five years. Don't worry if your goals change; this is a working document that can be edited and updated at any time.

4. Resume: Include a printed copy of your resume. If it is available online, include a link as well.

5. Skills: List 3-5 of your skills that you feel are relevant to your goals. For this section, try to include letters of recommendation and work samples to show off these skills. You can include classroom projects, certificates, and awards.

Working Document

Remember that since this is a working document, it should continually be updated. If you take any courses or do any volunteer work, be sure to add the information to your portfolio. You want to include anything that will make you stand out amongst all of the applicants!

Week 9 - PSAT/NMSQT Prep

- Math
- Evidence Based Writing & Language

https://www.lumoslearning.com/a/slh9-10

See Page 7 for Signup details

Weekly Fun Summer Photo Contest

 Take a picture of your summer fun activity and share it on Twitter or Instagram

Use the **#SummerLearning** mention

 @LumosLearning on Twitter or

@lumos.learning on Instagram

Tag friends and increase your chances of winning the contest

Participate and stand a chance to WIN $50 Amazon gift card!

Lumos Short Story Competition 2022

Write a short story based on your summer experiences and get a chance to win $100 cash prize + 1 year free subscription to Lumos StepUp + trophy with a certificate.
To enter the competition follow the instructions.

Step 1

Visit **www.lumoslearning.com/a/tg9-10**
and register for online fun summer program.

Step 2

After registration, your child can upload their summer story by logging into the student portal and clicking on **Lumos Short Story Competition 2022.**

Note: *If you have already registered this book and using online resources need not register again. Students can simply log in to the student portal and submit their story for the competition.*
Visit: www.lumoslearning.com/a/slh2022 for more information

Last date for submission is August 31, 2022

Use the space provided below for scratch work before uploading your summer story Scratch Work

2021 Winning Story

In March 2020, I found out that my 7th-grade exams were canceled. At first, I was excited, but I soon realized that these changes would upend my expectations for school. Over time, my classmates and I realized that the global coronavirus pandemic was not something to be excited about and would have long-lasting effects on our education. My school canceled exams again this year, and, strangely, I found myself missing them. The virus has revealed global inequality regarding health.

Even as America fights the virus, so is it also fighting racism and injustice. The Black Lives Matter movement has shown me how brutal racism can be. The deaths of George Floyd and Breonna Taylor, two African Americans killed by police for no reason, have made me aware of the dangerous injustice in America. Hatred and violence against Asian immigrants are also on the rise. People of color in the US are routinely subjected to prejudice, if not also violence, at the hands of white people. Chinese people are blamed for the "China virus,"; which has led to Asian Americans being attacked. Enduring forms of racism are preventing progress around the world. Racism in society takes many forms, including prejudice, discrimination, and microaggressions. If racism is systemic in America, there will never be true peace or equality until it is uprooted. People see me as a person of color and assume that I'm from Africa because of the color of my skin, even though I am half Black and half white. I don't seem to earn as much respect as a white person would because I am thought of as a foreigner, not a true American. It makes me feel unwelcome and unwanted. I am lucky to have access to technology to keep me engaged in learning. There are still others who don't have the ability to continue learning, whose educational institutions have been shut down by the virus. I have learned that so many people lack access to basic necessities and that racism in America continues to lead to violence and injustice. I aspire to work toward a system that addresses these inequalities in the future. This summer I reflected back on all these things and have learned that no matter what, we all should continue to push on, even through hardships and obstacles.

Submit Your Story Online & WIN Prizes!!!

2020 Winning Story

Finding Fun during a Pandemic

This was a weird summer. We did not travel because of COVID-19 and stayed mostly at home and outside around our house. Even when I saw my friends, it was unusual. This summer, I worked and made money helping my parents.

The pandemic allowed me to spend more time inside and I learned many new skills. We made face masks and had to figure out which pattern fits us the best. My sister and I enjoyed creating other arts and crafts projects. Additionally, I have been learning to play instruments such as the piano, guitar, and trombone. We also baked and cooked because we did not go out to eat (at all!). I love baking desserts. The brownies and cookies we made were amazing! I also read for one hour a day and did a workbook by Lumos Learning. I especially loved Math.

Our time outdoors was different this summer. We ordered hens. My family spent a lot of time fixing the coop and setting it up for our 18 chickens. We had a daily responsibility to take care of our chickens in the morning, giving them food and water and in the evening, securing them in their coop. We were surprised that 3 of the hens were actually roosters! Additionally, we exhausted many days gardening and building a retaining wall. Our garden has many different fruits and vegetables. The retaining wall required many heavy bricks, shoveling rocks, and moving dirt around. To cool off from doing all this hard work, we jumped in a stream and went tubing. Our dog, Coco liked to join us.

COVID-19 has also caused me to interact differently with my friends. We used FaceTime, Zoom, and Messenger Kids to chat and video talk with each other. Video chatting is not as fun as being in person with my friends. I love Messenger Kids because it is fun and you can play interactive games with each other.

I had to spend some of my time working. I helped clean my parents' Airbnb. This was busier because of COVID-19. My sister and I will start to sell the chicken eggs once they start to lay which we expect to happen anytime. We had a small business two years ago doing this same thing.

Summer 2020 has been unusual in many ways. We played indoors and outdoors at our house and nearby with family. I have learned new skills and learned to use technology in different ways. Summer of 2020 will never be forgotten!

Submit Your Story Online & WIN Prizes!!!

Answer Key
&
Detailed Explanations

Week 1 Summer Practice

Day 1

Question No.	Answer	Detailed Explanations
1	25	$125^{\frac{2}{3}}=(\sqrt[3]{125})^2 = 5^2 = 5\times5 = 25$. In a problem with a rational exponent, the numerator tells you the power, and the denominator the root. In the problem $125^{\frac{2}{3}}$, the denominator is 3 so you would take the cube root of 125 which equals 5. The numerator is 2 so you raise 5 to the power of 2 (i.e., 5×5) and the answer is 25.
2	B	$16^{\frac{5}{4}}=(\sqrt[4]{16})^5 = 2^5 = 2\times2\times2\times2\times2 = 32$. In a problem with a rational exponent, the numerator tells you the power, and the denominator the root. In the problem $16^{\frac{5}{4}}$, the denominator is 4 so you would take the fourth root of 16 which equals 2. The numerator is 5 so you raise 2 to the power of 5 (i.e., $2 \times 2 \times 2 \times 2 \times 2$) and the answer is 32.
3	B	The expression $\sqrt[a]{x^b}$ can be rewritten as $(x^b)^{\frac{1}{a}}$. This expression can be changed using the power of "a" power exponent rule, which states that when we raise an expression with "a" power to "a" power, we multiply the two exponents. Therefore, $(x^b)^{\frac{1}{a}}=x^{b\times\frac{1}{a}}=x^{\frac{b}{a}}$. Thus, $\sqrt[4]{x^3}$ $=(x^3)^{\frac{1}{4}}=x^{3\times\frac{1}{4}}=x^{\frac{3}{4}}$.
4	A	The rational exponent expression $x^{\frac{a}{b}}$ can be written as $x^{a\times\frac{1}{b}}$ by separating the numerator and denominator in the exponent. Next, the expression $x^{a\times\frac{1}{b}}$ can be written as $(x^a)^{\frac{1}{b}}$ using the power of "a" power exponent rule. Recall that $x^{\frac{1}{b}}$ is the equivalent of $\sqrt[b]{x}$, using the rational exponent rule. Now, we can put these rules together and get $y^{\frac{4}{5}} = y^{4\times\frac{1}{5}} = (y^4)^{\frac{1}{5}} = \sqrt[5]{y^4}$.

Question No.	Answer	Detailed Explanations
5	B	$3\sqrt{18} + 2\sqrt{2} = 3\sqrt{9\times2} + 2\sqrt{2} = 3 \times \sqrt{9} \times \sqrt{2} + 2\sqrt{2} = 3 \times 3 \times \sqrt{2} + 2\sqrt{2} = 9\sqrt{2} + 2\sqrt{2} = 11\sqrt{2}$. At first glance you might think this problem cannot be simplified because the radicands are 18 and 2, which are not the same. However, $\sqrt{18}$ is not simplified completely because it has a perfect square factor. The square root of 18 can be written as $\sqrt{9\times2}$. Then it can be written separately as $\sqrt{9} \times \sqrt{2}$. The square root of 9 is 3, so we can rewrite this term again as $3\sqrt{2}$. This makes our problem $3 \times 3\sqrt{2} + 2\sqrt{2}$. To keep simplifying we would multiply the 3's at the front of the problem resulting in $9\sqrt{2}+2\sqrt{2}$. Now we have simplified until our radicands are the same so we can add the coefficients 9 and 2, and keep the $\sqrt{2}$ the same. This gives a final answer of $11\sqrt{2}$.
6	B	The correct answer is B, because the feelings of mourning in the captain has almost made him cry. These show his feelings.
7	B	The correct answer is B because the word "annual" means that the two men go to the restaurant every year.
8		The correct evidence would be "When we were finishing our house, we found we had a little cash left over on account of the plumber not knowing it," or "I was for enlarging the hearth with it because I was always unaccountably down on the hearth somehow." These quotes from the passage reveal that Mr. McWilliams already has paid for another problem in the house and he doesn't want to pay for anything else. Also, he wants to pay for a hearth.
9		The old man didn't expect the demons to be there is based on the following line from the passage "He at once thought that his friends had come to look for him." The old man expected his friends and thought that is what he heard, so he didn't expect to see demons instead.
10	D	The correct answer is D because he is upset with the lump in the other choices, but the audience is made aware that he will do something about it by the last line of the paragraph.

Question No.	Answer	Detailed Explanations
1	82°F	To convert degrees Celsius to degrees Fahrenheit, you must use the formula given in the problem, $F=\frac{9}{5}C+32$. You would put in 28 for the C in the equation like this: $F=\frac{9}{5}(28)+32=82.4$. After doing the indicated math you get 82.4 and this would round down to 82. (Reminder: The rules for rounding say if the number after the decimal is 4 or below, you keep the number before the decimal the same when rounding. However, if the number after the decimal had been a 5 or more, we would need to round up to 83.)
2	1.05 hours per day	320 pages × 275 words per page = 88,000 total words → $\frac{88,000 \text{ total words}}{200 \text{ words per minute}}=440\text{ min} \rightarrow \frac{440 \text{ minutes}}{60 \text{ min per hour}}=7.33\text{ hours} \rightarrow \frac{7.33 \text{ hours}}{7 \text{ days}}$ =1.05 hours per day. You would need to multiply the number of pages by the words per page to get the total number of words that Alex needs to read. Then we can take that total number of words and divide it by the amount of words he reads per minute to get the total amount of minutes he will spend reading. Once you know the amount of time in minutes he will spend reading you can divide by 60 to convert the time into hours that he will spend reading. The final step is to divide the number of hours Alex will spend reading by the number of days he has to complete the reading assignment. This gives you the final result of 1.05 hours per day.
3	$3078	The problem states that the home has 1800 square feet. Since the price of the carpet is per square yard, we will convert the area inside the home to square yards. A yard is 3 feet, so a square yard is 3 feet by 3 feet or 9 square feet. Therefore, the home is 1800÷9=200 square yards. Multiply this quantity by the price per square yard: 200×15.39= 3078.
4	B	$1216\times\frac{1}{3}=405.33$. To find the answer you would multiply the number of students in your high school by one-third. You get a decimal answer and since it doesn't make sense for there to be .33 of a person you round the exact answer to 405 votes. This would be correct if EVERYONE in your high school casts a vote but the problem said only that the majority of the students voted... not all of them. So we pick the closest number to 405 without going over and get 398 as an acceptable answer.

Question No.	Answer	Detailed Explanations
5	D	Answer choice A would not work because his friend did not travel during the same time of year Jacob will be traveling to Hawaii. Finding the current ticket price will not work because Jacob is not going to be traveling right now so answer choice B is also out. Asking his mom, who went to Florida last summer, would not help for many reasons. The most important reason being she did not go to the same destination he is planning to go to. The only answer that makes any sense at all to help get an estimate of the plane ticket cost is choice D.
6	C	The correct answer is C because each man is described and they worked together under the command of the captain, so they worked as a team.
7	A	The correct answer is A because the two men ate a lot of dinner for Thanksgiving and it resulted in trips to the hospital.
8		The man finds happiness after being depressed for so long and the two stories come together with the demons providing the old man with happiness. The old man tried to get the lump removed or do something about it, but he ended up finding happiness even though his lump was still on his face.
9	B	The correct answer is B because the Cyclops are described as being savages and that they should be feared.
10	C	The correct answer is C. The narrator admits that he is mad and deeply in love with Eleanor.

Question No.	Answer	Detailed Explanations
1	C	The degree of a term is the sum of the exponents of the variables in the term. The degrees of the terms are in the table below.

Term	Degree
$7x^9y$	10
$-9xy^8$	9
$2x^5y^6$	11
$-5xy$	2

Question No.	Answer	Detailed Explanations
2	B	A factor is a number that when multiplied gives another number or an expression. The given expression $36x^2+12x+24$ can be factored as $12(3x^2+x+2)$ which shows that 12 is a factor of the given expression.
3	D	The expression $5x^3y^4+7x^2y^3-6xy^2-8xy$ is a polynomial expression with four terms. The coefficient of a term is the number in the front of the term. If the term begins with a negative, then the coefficient is a negative number, whether or not the term has variables. The third term is $-6xy^2$ and the number in the front of the term is -6.
4	8	A factor is a number that when multiplied gives another number or an expression. The given expression $16x^4-8x+64$ can be factored as $8(2x^4-x+8)$ which shows that 8 is a factor of the given expression.
5	B	The given quadratic function, $f(x)=9x^2+66x+21$, has a greatest common factor (GCF) of 3. Start by factoring out the GCF: $f(x)=3(3x^2+22x+7)$. Next, factor the first and third terms so the factors give the middle coefficient when the binomials are multiplied using the FOIL process. Since all terms are positive, we know that all factors will have positive numbers. $f(x)=3(3x^2+22x+7) = 3(3x+1)(x+7)$. Now, set the function equal to zero and use the zero product property to find the zeros of the function. $3(3x+1)(x+7)=0$; $3x+1=0$; $x+7=0$; Solving the two equations, we find that $x=-\dfrac{1}{3}$; $x=-7$.

Question No.	Answer	Detailed Explanations
6	A	The correct answer is A because the four men know that they must work as a team to get through this ordeal.
7	C	The correct answer is C because Hermia would most likely become fearful that her father would resort to kill her for disobeying him.
8	B	The correct answer is B because the waiters make a comment that shows they don't think the Old Gentleman should eat at the restaurant with Stuffy.
9		Since Mr. McWilliams does not believe in burglar alarms, he almost expects burglars to come in. When the burglar does come in, Mr. McWilliams speaks with him in a humorous and respectful way because Mr. McWilliams expecting the burglar to come in. Mr. McWilliams's expectations contribute to the theme because they are the same. They both expect the burglar alarms not to work and be invaluable. Therefore, when the burglar comes into the house, Mr. McWilliams is not surprised.
10		

	calm and relaxed (A)	greedy (B)	closed-minded (C)	fearful (D)
Mr. McWilliams			●	
Mrs. McWilliams				●
Burglar	●			
Burlgar Alarm man		●		

The correct answer is Mr. McWilliams is closed-minded (C) about the burglar alarm because he explains how he doesn't think they work well.

The correct answer is Mrs. McWilliams is fearful of burglars (D) because she wants a burglar alarm and makes her husband get one for her.

The correct answer is The burglar acts calm and relaxed (A) because he is smoking a cigar and doesn't seem concerned.

The correct answer is The burglar alarm man who installs and fixes the alarm is (B) because he charges the family every time he comes out.

Day 4

Question No.	Answer	Detailed Explanations
1	C	If we divide any term by its previous term, we will get the common ratio r of the series. So, $r=\frac{10}{2}=5$
2	A	The question states that the finite geometric series has $a_1=6$, $a_2=2$, $a_3=\frac{2}{3}$, $a_4=\frac{2}{9}$. If we divide any term by its previous term, we will find the common ratio r in the series. Thus, $r=\frac{2}{6}=\frac{1}{3}$. The general formula for the nth term of a geometric series is $a_n=a_1 \times r^{n-1}$ Therefore, $a_n=6.(\frac{1}{3})^{n-1}$
3	D	When combining polynomials, combine like terms by combining the coefficients. $(9x^3+2x^2-4x+1) + (-5x^3-x^2-5x+7)$ $(9x^3-5x^3) + (2x^2-x^2) + (-4x-5x) + (1+7)$ $4x^3+ x^2 -9x + 8$
4	B	When combining polynomials, combine like terms by combining the coefficients. $(2x^3 + 5x^2 - 7x - 8) - (7x^3 - 8x^2 + 5x - 3)$ $(2x^3 - 7x^3) + (5x^2 - (- 8x^2)) + (- 7x - 5x) + (- 8 + 3)$ $- 5x^3 + 13x^2 - 12x - 5$
5	D	Substitute $x = 1$ and solve; if the substitution yields zero, there is no remainder; in this case $2(1)^3-5(1)+(1)-3=-5$.
6	B	The correct answer is B because the situation is serious and a bucking bronco would show how violent the waves are pounding into the dinghy.
7	B	The correct answer is B because it contains more formal old English words that were used and spoken in a more proper way during Shakespeare's time.
8	C	The correct answer is C because the author is being funny when he describes Stuffy going to the hospital.
9	A	The correct answer is A because the men wanted to eat a lot and they stuffed themselves which is compared to them fighting against an enemy.
10	B	The correct answer is B because when the burglar took "a lot of miscellaneous property," he probably grabbed a bunch of things very fast without looking to see if they were valuable. The burglar did not go around and look carefully at what he would take.

Day 5

Question No.	Answer	Detailed Explanations
1	B	The given quadratic equation is $x^2 - x - 20 = 0$. Next, factor the first and third terms so the factors give the middle coefficient when the binomials are multiplied using the FOIL process. $x^2 - 5x + 4x - 20 = 0$ $x(x - 5) + 4(x - 5) = 0$ $(x - 5)(x + 4) = 0$ So, $x - 5 = 0$ or $x + 4 = 0$ Therefore, $x = 5$ or -4.
2	D	By rearranging the given equation $6x^2 - 18x - 18 = 6$ $6x^2 - 18x - 18 - 6 = 0$ $6x^2 - 18x - 24 = 0$. Here, the quadratic equation $6x^2 - 18x - 24 = 0$ has a greatest common factor of 6. Divide the equation by 6 : $x^2 - 3x - 4 = 0$ Next, factor the first and third terms so the factors give the middle coefficient when the binomials are multiplied using the FOIL process. $x^2 - 4x + 1x - 4 = 0$ $x(x - 4) + 1(x - 4) = 0$ $(x - 4)(x + 1) = 0$ So, $x - 4 = 0$ or $x + 1 = 0$ Therefore, $x = 4$ or -1.
3	A	Student may graph this function to see that it has only one x-intercept at $(1, 0)$.
4	A	Student may use The Remainder Theorem and synthetic division or simply substitute -3 into the polynomial expression to find the value of the function at -3, student must be careful with the negative sign when substituting. $f(-3) = (-3)^3 + 3(-3)^2 + 2(-3) + 9$ $\quad\quad = -27 + 27 - 6 + 9$ $\quad\quad = 3$

Question No.	Answer	Detailed Explanations
5	B	Student may use The Remainder Theorem and synthetic division or simply substitute 2 into the polynomial expression to find the value of the function at 2, student must be careful with the negative sign when substituting. $x^3 - x^2 - x - 2$ $= 2^3 - 2^2 - 2 - 2$ $= 8 - 4 - 2 - 2$ $= 0.$ Here, The remainder is 0. Therefore, $(x - 2)$ is a factor of $x^3 - x^2 - x - 2$.
6	B	The correct answer is B. Denotative is the literal, or exact meaning of a word without any shades of meaning.
7	A	The correct answer is A because the connotative definition of a word is designed to invoke an emotional response to the word
8	B	The connotation of a word is designed to provoke an emotional response, so based on our emotional response to the word, we can analyze the tone of the piece or how the narrator sounds.
9		Dickenson is comparing hope to a bird in this line, and the metaphor is that hope is like a bird that sits in one's soul.
10	True	The answer is true, the words "sore" and "abash" allow the reader to understand the strength of the storm.

Question No.	Answer	Detailed Explanations
1	C	Student must use the formula for binomial expansion to simplify and find the coefficients. By using the formula for binomial expansion $(r + 2)^3$ $= (r)^3 + 3 \times (r)^2 \times (2) + 3 \times (r) \times (2)^2 + (2)^3$ $= r^3 + 6r^2 + 12r + 8$
2	A	Student must use the binomial expansion formula as follows: $= 5C_0 a^5 b^0 = a^5$
3	B	Student must factor the binomial numerator using the GCF of 3 to get 3(s - 2). So, $\dfrac{3s - 6}{s - 2} = \dfrac{3(s - 2)}{s - 2} = 3.$
4	C	Student must factor the Numerator and denominator separately, $\dfrac{x^2+8x+12}{x^2+3x-18}$ $\dfrac{(x+2)(x+6)}{(x-3)(x+6)}$ $= \dfrac{x+2}{x-3}$
5	B	Student must find that the LCM of 3 and 5 is 15 and the LCM of x and x is x.
6	B	The correct answer is B because tension is created as the author explains each movement of the wave causing problems on the dinghy and the men unsure if they will make it to shore safely.
7	A	The correct answer is A because Shakespeare's version includes the dialogue or what the characters are saying to each other.
8	A	The correct answer is A because two separate stories exist after the dinner when Stuffy was taken to the hospital and the Old Gentleman was taken to the hospital.

Question No.	Answer	Detailed Explanations
9	D	The correct answer is D because a situation occurs that is not expected to happen.
10		There are a few hints of foreshadow that the McWilliams will have trouble with burglars. First, Mr. McWilliams lets it be known at the beginning of the story that he does not see the value in the burglar alarms. Second, the burglar alarm man says that this burglar alarm only occurs on the first floor. When the burglar alarm man says that the alarm only works on the first floor, it is expected that another burglar will come until each floor contains the alarm.

Question No.	Answer	Detailed Explanations
1	C	Student must find the LCD as 2b and then adjust the numerators accordingly, lastly realize that the numerators are unlike terms and cannot be simplified therefore are written as a binomial. $$\frac{a}{b} - \frac{1}{2}$$ $$= \frac{a}{b} \times \frac{2}{2} - \frac{1}{2} \times \frac{b}{b}$$ $$= \frac{2a}{2b} - \frac{b}{2b}$$ $$= \frac{2a-b}{2b}$$
2	C	Student must factor all terms in the numerators and denominators and then simplify. $$\frac{x^2-1}{x+2} \times \frac{x+2}{2x-2}$$ $$= \frac{(x+1)(x-1)}{(x+2)} \times \frac{(x+2)}{2(x-1)}$$ $$= \frac{x+1}{2}$$
3	No	In order to be a function, each x-value should be paired with exactly one y-value. Since we have the x-value -1 paired with both 5 and 2, this is not a function.
4		For any ordered pairs of a function are given as (x_1,y_1), (x_2,y_2), (x_3,y_3), (x_4,y_4). Then the domain of a function is (x_1, x_2, x_3, x_4) and range of a function is (y_1, y_2, y_3, y_4). So, here the domain is $(-5,-4,-3,-2,-1)$.
5	D	The input variable is x. Doubling the square of the input variable is a term $2x^2$. Subtracting 13 gives us the term -13. Put together, we have $g(x)=2x^2-13$. Therefore, $g(3)=2(3)^2-13=18-13=5$.
6	B	The correct answer is B because as a news reporter, who was not part of the ship's crew, was an average man on the ship.
7	D	The correct answer is D because this culture reveals that women were not given the same rights as men due to their father being able kill their daughter for not marrying the man of his choice.

Question No.	Answer	Detailed Explanations
8		The narrator's point of view concerning the effectiveness of the burglar alarm can be described as worthless and costing too much money for how much he has to pay. This is because the husband shows signs that he really doesn't think that burglar alarms work and therefore are not worth the money.
9	D	The correct answer is D because in today's time, usually there are not coachmen because people don't ride horses for transportation purposes.
10		Based on the man's view of the demons, his point of view of the demons is that they are not threatening to him and are enjoying themselves.
		The demons are dancing and having fun which makes the old man enjoy watching them and he doesn't feel scared of them.

Week 2

Question No.	Answer	Detailed Explanations
1	34	In order to find the next number in the Fibonacci sequence we need to add the two previous terms. We are given the first 7 terms, (1, 1, 2, 3, 5, 8, 13....) so to find the 8th term we need to add the 6th and 7th terms and get $8 + 13 = 21$. Then to find the 9th term we need to add the 7th and 8th terms and get $13 + 21 = 34$. So the 9th term in the Fibonacci sequence is 34.
2	7 and 9	In this sequence, we get the next number by adding 2 to the previous number. So to get the 4th number we add 2 to the 3th number and get $5 + 2 = 7$. Then to get the 5th number we add 2 to the 4th number and get $7 + 2 = 9$. So the next two numbers in this sequence should have been 7 and 9.
3	-3||1	(see table and work below)
4	C	(see explanation below)

Question 3 table:

x	f(x)
-3	0
-2	-3
-1	-4
0	-3
1	0
2	5

You can find these values easily using the table function on the graphing calculator. You can also factor and solve to find the x-intercepts as shown below.

$x^2 + 2x - 3 = 0$
$(x + 3)(x - 1) = 0$
$x + 3 = 0$ and $x - 1 = 0$
$x = -3$ and $x = 1$

Question 4 explanation:

Since the y-intercept occurs at (0, 6) we know that it has to be answer choice A or C because the c value in standard form $ax^2 + bx + c$ is the y-intercept. Therefore the equation must have a "+6" at the end.

You can use the x-intercepts and work backwards to find the rest of the equation.

Since the zeros are -2 and -3, we can write factors $(x+2)(x+3)$. Now use the FOIL method and multiply those together and arrive at $x^2 + 5x + 6$ which is answer choice C.

Question No.	Answer	Detailed Explanations
5	B	Since dad placed a cap of $20, you can only buy up to $\frac{20}{0.50} = 40$ songs per month. The smallest number of songs you can buy per month is zero. Therefore the domain for this problem is between 0 and 40 as stated in answer choice B.
6	D	The correct answer is D because both the story shows how the men are struggling with the ocean and the letter compares the sea to a stage.
7	A	The correct answer is A because a large portion of this excerpt is given to the father, Egeus, to explain why he brought his daughter to court.
8		The itemized list of materials used to secure the McWilliams' property shows that they had to ask the expert several times and either he didn't live close by or the parts he needed had to be shipped in. This is shown by having two sets of labor that Mr. McWilliams had to pay. Plus, there is a railroad charge, so the expert had to come from far away to secure their house.
9	A	The correct answer is A because the list is detailed so that the reader will sympathize that the McWilliamses are being taken advantage of with all of the costs.
10	B	The correct answer is B because they both describe how the Cyclops lived off of the land by planting which refers to their economy.

Week 2

Day 4

Question No.	Answer	Detailed Explanations
1	31.5 pages per day	To find the rate, we simply divide $\frac{126}{4}=31.5$, so Jacob read about 31.5 pages per day.
2	No	Damon's rate of change was $\frac{126}{2}=60$mph and Ashley's rate of change was $\frac{135}{2.3}=58.7$mph. So Damon's rate was actually greater than Ashley's.
3		

x	f(x)
-2	4
-1	1
0	0
1	1
2	4

The quadratic parent function is $y=x^2$. To get the missing value simply plug $x=2$.
$y=x^2=2^2=4$

Question No.	Answer	Detailed Explanations
4	B	The question asks us to select a function that is equivalent to $g(x)=x^4+x^8-x^{10}$. The terms in this function have a greatest common factor (GCF) of x^4. We can factor the GCF out of the function, divide each term by the GCF, and still have an equivalent function: $g(x)=x^4+x^8-x^{10}=x^4(\frac{x^4}{x^4}+\frac{x^8}{x^4}-\frac{x^{10}}{x^4})=x^4(1+x^4-x^6)$
5	C	Linear functions graphs will be straight lines. The only graph that is a straight line (with no curve at all) is C.
6	B	The correct answer is B because the Lamb's version describes Egeus as "stern" which makes him seem more like a human being and as a concerned father. In Shakespeare's version, he is described that he should be treated as a god.
7	B	The correct answer is B because the Lambs decided to explain the information about the law whereas Shakespeare relayed the information to the reader by the characters' words.

Question No.	Answer	Detailed Explanations
8		

x	Havell's version	Homer's original version
Created a sense of fear when Odysseus arrived	◯	
Portrays the Cyclops as a destructive force		◯
Described the lack of government where the Cyclops lived		◯

Havell's version created a sense of fear when Odysseus arrived because Havell discusses that they came to the land on a dark night with a mist forming. Homer shows that Cyclops was a destructive force by describing them as a "savage kind." Finally, Homer's version explains that the Cyclops did not have a government because they didn't have a monarch.

Question No.	Answer	Detailed Explanations
9	C	The correct answer is C because Homer's version describes that they made wine from the grape clusters.
10		The way that the Cyclops did not interact with those around them is shown in the original version by describing that they did not associate with their neighbors.

In the original version, it shows that "each rules his race." This means that these creatures or people kept to themselves. |

Question No.	Answer	Detailed Explanations
1	A	To calculate the rate of growth of the functions, we calculate their slope. The slope between two points (x_1,y_1) and (x_2,y_2) is calculated with the slope formula: $\frac{y_2 \cdot y_1}{x_2 \cdot x_1}$. In this problem, we estimate the slope of f(x) with the two points (4,8) and (8,32) so, using the formula, we have $\frac{32-88}{8-4}=\frac{24}{4}=6$. The two points for g(x) are (0,1) and (8,4) so, using the formula, we have $\frac{4-1}{8-0}=\frac{3}{8}$. Therefore, f(x) has a faster growth rate at x=6.
2	C	To calculate the rate of growth of the functions, we calculate their slope. The slope between two points (x_1,y_1) and (x_2,y_2) is calculated with the slope formula: $\frac{y_2-y_1}{x_2-x_1}$. In this problem, we estimate the slope of h(x) with the two points (3,1) and (9,9) so, using the formula, we have $\frac{9-1}{9-3}=\frac{8}{6}=\frac{4}{3}$. The function k(x) is a linear function in slope-intercept form y=mx+b. The slope of k(x) is 2. Therefore, k(x) has a faster growth rate between x=3 and x = 9.
3	C	The question states that one dollar is invested in an account that accumulates 20% interest per year. This means that the value of the account increases proportionally based on the amount in the account already. The formula for this type of function is f(x) = 1(1.2)^x which has a growth factor of 1.2 per year. A graph of this function is below.

Question No.	Answer	Detailed Explanations
4	A	The question states that the temperature began at 2°C and decreases 5°C each hour. This means that the beginning temperature is the y-intercept and the temperature decreases the same amount (5°C) each hour, which means the slope of the graph is -5. This situation can be represented by the linear function $f(x) = -5x+2$. A table of values of this function is below.

x	-4	-3	-2	-1
y	22	17	12	7

Question No.	Answer	Detailed Explanations
5	C	The question states that the pressure inside a chamber of a compressed oxygen tank begins at -2 PSI and every day, the pressure decreases 20% more than the previous day. This means that the pressure in the chamber decreases proportionally based on the previous days pressure. The formula for this type of function is $f(x) = -2(1.2)^x$ which has a growth factor of 1.2 per day, and even though it is in the negative direction, the function is an exponential growth function. A graph of this function is below.

Question No.	Answer	Detailed Explanations
6	B	The correct answer is B. This is called an adaption of a piece of literature, when you transform an original text into another setting or style.
7	D	The answer is D. The adapted characters are the pigs and the wolf. The theme remains the same (protecting oneself), and the antagonist is still the wolf as evident by the words huff and puff.
8		Huff and puff
9		In this version, the pig uses fiberglass which is cheap and readily available, like the hay in the original version.
10	A	The correct answer is A, setting. The writer has changed the time and place of the story to a more futuristic setting.

Question No.	Answer	Detailed Explanations
1	A	The graph shows a linear function, so the formula, in slope-intercept form is $f(x)=mx+b$, where m is the slope of the line (rise over run) and b is the y-intercept. Looking at the graph, select the two points $(4,5)$ and $(-16,0)$ to calculate the slope. The formula and calculations are: $m=\dfrac{y_2-y_1}{x_2-x_1}=\dfrac{5-4}{4-1}=14$ The graph shows that the y-intercept is $(0,4)$, so $b=4$. The function whose graph is shown is $f(x)=\dfrac{x}{4}+4$.
2	A	The questions states that the sequence is arithmetic. The formula for an arithmetic sequence is $a_n=a_1+d(n-1)$, where a_n is the nth term, a_1 is the first term, and d is the difference between each term. The table shows that $a_1=5$ and each term is 5 more than the previous term so $d=5$. Using these numbers, the sequence rule is $a_n=5+5(n-1)=5+5n-5=5n$.
3	C	Carefully compare the values in the graph and in the table. The y-values in the graph and the y-values in the table are approximately the same in the interval [5,6]. Then, to confirm this information, determine the functions represented. The graph represents $f(x)=(1.2)^x$. The table represents the function $f(x)=0.15x+2$. Using a graphing calculator, find that these two functions intersect at the point $(5.78,2.87)$. (Values are rounded to the nearest two decimal places.) After that point, the exponential function has larger values than the linear function.

Question No.	Answer	Detailed Explanations
4	D	Carefully compare the values in the graph and in the table. The y-values in the graph and the y-values in the table are approximately the same in the interval [3,4]. Then, to confirm this information, determine the functions represented. The graph represents $f(x)=\frac{1}{3}(x-2)^2-2$. The table represents the function $f(x)=\frac{1}{2}x-2$. Using a graphing calculator, find that these two functions intersect at the point $(3.5,-0.25)$. After that point, the quadratic function has larger values than the linear function.
5	A	If nothing is written at the base of the log, we should always assume a base of 10. The simple reason why is tradition.
6	B	The correct answer is B because the Chicago packaging market used a similar process for packaging meat.
7	D	The correct answer is D because if the process is used everywhere, then it would be very popular.

8

	"The strength of the Indian's bow and arrow is shown compared to that of others."	"Supplies are specific for optimum results."	"Techniques are described to keep the arrows strong."
After this the arrow should be painted, both to keep it from warping and to make it easier to find in the forest by its bright color.			⭕
We hear that many times Indians shot so hard that their arrows appeared on the far side of the animal, but the long bow, such was used by the old English archers or bowmen, was much the more powerful.	⭕		
These feathers are best made from either turkey or goose wings.		⭕	

It describes that the arrow should be painted to keep the arrow strong and prevent it from warping.

The second evidence shows that the Indian's bow and arrow were much better than others.

The third evidence shows that specific feathers should be used for the best arrows.

Question No.	Answer	Detailed Explanations
9	B	The correct answer is B because this means that there are many other functions that could not be described.
10	A	The correct answer is A because this mentions the sperm whale and that it is a toothed whale unlike the other two.

Question No.	Answer	Detailed Explanations
1	A	The number of hamburgers in the cooler is calculated by subtracting the number of hamburgers sold from the beginning inventory of hamburgers. The problem states that the restaurant has 1,090 hamburgers at the beginning of the week. The number of hamburgers sold is calculated by multiplying the sales per day by the number of days. The problem also states that the restaurant business expects to sell 150 hamburgers every day and the variable p represents the number of days. Therefore, the function that represents the number of hamburgers in the cooler after p days is $H(p)=1090-150p$.
2	B	The company has 150 sheets on hand and receives t truckloads with 80 sheets in each truckload. Hence, the total number of sheets the company has will be $P(t) = 150 + 80t$.
3	C	Answer is C. The definition is given in the question for a line segment. Angles and circles are not defined by lines, and perpendicular lines involve a pair of lines that are not bounded.
4	A	Answer is A. A ray is defined as having one end point and going an infinite length from that end point. A beam of light starts at it's source and continues infinitely away from it into the sky. Although the other examples seem to have no end, there are starting and ending points, or with the example of the equator, it makes a circle.
5	D	Answer is D. Translations slide the point in the given amount, shown in the graph below. 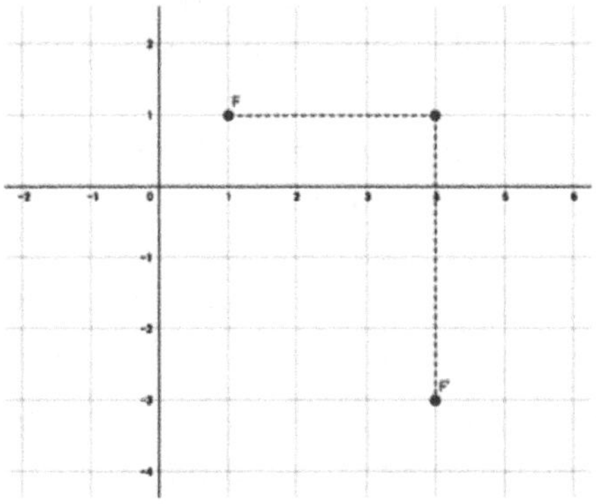

Question No.	Answer	Detailed Explanations

6

	Society	Government
Brings together what we like	◯	
Holds in our evil		◯
Created by our desires	◯	
Produced by mischief		◯

Paine promotes society and talks positively about society, but negatively about the government.

7

The theme that is addressed in "Common Sense" refers to the need for a new government due to the problems it is bringing onto the colonists.

This theme can be shown in Paine's comparison of society and the government and the example he provides.

8   **A**

The correct answer is A because this detail supports the central idea about the effectiveness of the assembly line. This shows how fast the process can go.

9

	Disagreement of needed regulations	Did not use parts of Parliament bodies	Useless meetings
dissolved Representative Houses repeatedly		◯	
refused his Assent to Laws	◯		
called together legislative bodies at places unusual, uncomfortable, and distant			◯

10

The central idea emerges through the description of the telegraph's parts and what it does to receive or transmit the message.

Each part of the telegraph is explained along with how it functions in order to produce the entire message.

Day 3

Question No.	Answer	Detailed Explanations
1	C	Answer is C. The reflection over x = 1 is able to match the shapes, however it needs to be translated, or slide two units down in order to map onto itself.
2	D	Answer is D. The image needs to slide to the right four units, positive 4 on the x-axis, and down two units, negative 2 on the y-axis.
3	B	Answer is B. A translation slides, or moves the shape by adding or subtracting from the coordinates.
4	A	A translation will move, or slide the shape. In this case all points go three to the left and two down.
5	C	All points are equidistant to the line x = -2
6		In the third paragraph, the author includes an example to describe how the government would want to be built by the citizens themselves. Paine discusses how a group of people would function on an island in regard to creating a governmental system.
7		Thomas Paine emphasizes that the government's purpose is to steal from the citizens in order to make sure the leaders are rich. This is clearly shown in the first two paragraphs when it is comparing the government to society.
8		The author does unfold the information about his assembly line by providing specific facts about each step of the process. Since the author used this process, he describes it in great detail.
9	A	The correct answer is A because the author explains the materials as the author is describing each part of the process.
10		The bow and arrow are introduced in the passage by describing how powerful the Indian's bow and arrow was compared to others and the boys today should learn how to make and shoot them. This provides an introduction so that the reader will sense the importance of this bow and arrow as opposed to others.

Week 3

Question No.	Answer	Detailed Explanations
1	A & C	The image is not regular and can be reflected over the line of symmetry and rotation of 180° only.
2	B	A stretch will change the lengths and angle measures of a shape and the dilation will increase or decrease the lengths of the sides of a shape.
3	A, B, C, D	Shapes need to have the lengths of the sides equivalent as well as the angle measures. If these are equal then the area and perimeters are also.
4	D	By having the ladder slide over, it is a translation. Since all of the angles and lengths are preserved, they are congruent
5	C	Only answer choice C matches with congruency postulate. In this case, it is the SAS rule gets applied. The sides and the angles must be incorrect order such as the corresponding side and included angle. Hence, answer choice C is correct.
6		The usage of "had adopted the policy of 'man-high' work" refers to that the men would be on their toes concentrating about their work at all times. The reason for this is because the men could not take a break whenever they wanted to because that would cause the others in line to slow down and have to wait.
7		

	A device for raising or lowering something	Electric generator	The framework on the car
fly-wheel magneto		◯	
chassis			◯
windlass	◯		

The fly-wheel magneto is an electric generator.
The word chassis means the framework of the car.
The windlass is a device for raising or lowering something.assis" has the definition stated in the passage.

Question No.	Answer	Detailed Explanations
8	D	The correct answer is D because the midrib is described in the sentence as a vein which is close to the meaning of the stem.
9		The ferrule is found on the top or bottom of an object.
10	B	The correct answer is B because a metropolis is a city like London.

Week 3

Question No.	Answer	Detailed Explanations
1	A & B	Answers are A and B. Vertical and Alternate Interior angles are congruent. Supplementary angles have a sum of 180 degrees and complementary angles have a sum of 90 degrees.
2	D	Answer is D. The sum of all of the interior angles in a triangle is 180. $180 - (47+38)=95$
3	B	Answer is B. It is the base angles of isosceles triangles theorem.
4	C	Answer is C. The interior angles of a triangle must have a sum of 180 degrees. $180 - (75 + 30) = 75$
5	B	Answer is B. A scalene triangle is defined by its side lengths, none of them are congruent.
6	A	One can assume it means large and powerful because the enemy appears so great judging by the urgency of his tone and the lengths he is going to persuade the people to fight.
7	A	The answer is A because sarcasm is the use of irony to emphasize disgust for a subject.
8	B	He wants his people to rise up and revolt against the British rule.
9	B	He uses B, metaphor, to compare life under British rule to slavery. He says by not fighting we purchase peace, but the price is chains. This is a comparison between two ideas.
10	C	The answer is C because both are true.

Week 4 Summer Practice

Day 1

Question No.	Answer	Detailed Explanations
1	C	Answer is C. A parallelogram has two pairs of parallel sides, a trapezoid only has one pair of parallel sides.
2	A	Answer is A. Opposite angles of a parallelogram are congruent
3	B	Answer is B. The definition of a regular polygon is that all sides, and angles, are equal.
4	Eric	The definition of inscribe is one shape resides in the other shape. The order in which the polygons are presented tells the order of which shape is inside. Since the triangle is inscribed in a circle, the triangle is inside of the circle.
5	B	The original circle F has its center at the point $(-6,6)$ with a radius of 4 units. The translated/dilated circle F' has its center at the point $(-2,-8)$ with a radius of 2 units. This means the center was translated right 4 units and down 14 units. As a transformation, this translation is written as $(x,y) \rightarrow (x+4, y-14)$. Circle F was also dilated by a factor of $\frac{1}{2}$ because the radius was reduced from 4 units to 2 units. As a transformation, this dilation is written as $(x,y) \rightarrow \frac{1}{2}(x,y)$. Putting the translation and dilation together, the rule is $(x,y) \rightarrow \frac{1}{2}(x+4, y-14)$.
6	B	The correct answer is B because Paine believes that the citizens should not be harmed and there is a difference between society and the government.
7	B	The correct answer is B because the author uses details to show step-by-step how to make the bow and arrow.
8		The author's claim about the importance of the telegraph is developed through the way it works as well as the importance of the machine. This allows the author to reveal how important the telegraph is to the rest of the world as well as how simple it is to use.

Question No.	Answer	Detailed Explanations
9		The author indirectly claims that different whales are difficult to tell apart. The author provides a lot of similarities among the whales and he explains how there are only certain body parts and its location to tell these animals apart.
10		The author explains that legally the flooding was caused by an "act of God" because no one is to blame for this act. An "act of God" is more difficult to prepare for since these natural disasters can't be predicted.

Question No.	Answer	Detailed Explanations
1	B	A corollary to the Inscribed Angle Theorem states that the measures of opposite angles in a quadrilateral that is inscribed in a circle are supplementary angles. The sum of the measures of supplementary angles is 180°. Quadrilateral ABCD is inscribed in circle O, and $\angle DAB$ is opposite $\angle BCD$, and $m\angle BCD=110°$, as shown in the figure. Therefore, $\angle DAB=70°$ because $110+70=180$.
2	A	A geometry theorem states that the measure of an inscribed angle is equal to half the measure of its intercepted arc. It follows, then, that the measure of an intercepted arc is two times the measure of the inscribed angle that forms the arc. According to the figure, $\angle VUW$ is an inscribed angle and forms the arc $\overparen{VW}$. Since $m\angle VUW=30°$, $m\overparen{VW}=60°$.
3	A	A geometry theorem states that the tangent segments from a point outside the circle to two tangent points on the circle are congruent. Thus, $\overline{VZ}\cong\overline{WZ}$, $\overline{WX}\cong\overline{UX}$, and $\overline{UY}\cong\overline{VY}$. Then, $m\overline{VZ}=m\overline{WZ}=4$, $m\overline{WX}=m\overline{UX}=8$, and $m\overline{UY}=m\overline{VY}=9$. Now, by the Segment Addition Postulate, $m\overline{XY}=8+9=17$.
4	A	A geometry theorem states that the tangent segments from a point outside the circle to two tangent points on the circle are congruent. Thus, $\overline{UX}\cong\overline{UZ}$, $\overline{VX}\cong\overline{VY}$, and $\overline{WY}\cong\overline{WZ}$. Then, $m\overline{UX}=m\overline{UZ}=6$, $m\overline{VX}=m\overline{VY}=5$, and $m\overline{WY}=m\overline{WZ}=5$. Now, by the Segment Addition Postulate, $m\overline{UW}=5+6=11$.
5	C	81A geometry theorem states that if a line is tangent to a circle at the outer point of a radius, then the radius is perpendicular to the tangent line. Thus, $\triangle OQR$ is a right triangle and the relationship between the sides and the hypotenuse must obey the Pythagorean Theorem. This means that $(OQ)^2+(QR)^2=(OR)^2$, so $x^2+15^2=(x+9)^2$. Solve the equation $x^2+15^2=(x+9)^2$. Using correct order of operations, perform the exponents in the equation, giving $x^2+225=x^2+18x+81$. Cancel the x^2 on each side, leaving $225=18x+81$. Subtract 81 from both sides and divide both sides by 18. The solution to the equation is $x=8$.
6		The purpose of Thomas Paine writing "Common Sense" was to cause the British government to change their laws or create a new government decided by the citizens in America. Paine describes how a group of people should be able to make their decisions regarding their government.

Question No.	Answer	Detailed Explanations
7	C	The correct answer is C because the author is proud the he can describe this new process for making cars.
8		The author includes this detail because the next man on the assembly line would have to unscrew or undo what the previous man had done.
9	A	The correct answer is A because Jefferson's bitterness and anger is shown by his accusations of the British government.
10		The purpose of the drawing is to show what the different parts of the Indian's bow and arrow look like. This image allows the reader to see the different parts of the bow and arrow as well as the detailed steps to make them.

Question No.	Answer	Detailed Explanations
1	C	The area of a circle is calculated using the formula $A=\pi r^2$, where A is the area and r is the radius. The area of a sector of a circle is calculated using the formula $A=\frac{b}{360}\times\pi r^2$, where b is the degrees of the sector of the circle. The figure shows that the radius is 12 yd and the sector has a measure of 235°. Thus, the approximate area is $A=\frac{235}{360}\times(3.14)(12)^2=295.16$ yd².
2	B	The circumference of a circle is the distance around the circle, and is calculated with the formula $C=2\pi r$. An arc is a part of the circumference and the length of an arc in a circle is calculated using the formula $L=\theta r$, where L is the length of the arc, θ is the measure of the angle in radians, and r is the radius. The figure shows that the radius is 14cm and the sector has a measure of 135°. Convert the angle to radians by multiplying by $\frac{\theta}{180°}$. Thus, the measure of the angle in radians is $135\times\frac{\pi}{180°}=\frac{3\pi}{4}$. Therefore, the length of the arc is $L=\frac{3\pi}{4}\times14$cm$=\frac{21}{2}\pi$ cm.
3	B	The equation of a circle is in the form $(x-h)^2+(y-k)^2=r^2$. The general form of the point that is the center of a circle is (h,k), so $h=-4$, and $k=-7$. This gives us the equation $(x+4)^2+(y+7)^2=r^2$. The question says the circle contains the point $(4,-9)$. This means that the radius of the circle is the distance between the points $(-4,-7)$ and$(4,-9)$. Use the distance formula to find the radius squared. $r=\sqrt{(-4-4)^2+(-7+9)^2}=\sqrt{68}$; so $r^2=68$. The equation of the circle is $(x+4)^2+(y+7)^2=68$. Now, substitute the points into the equation for x and y. If the result is a true statement, then the point is on the circle. Beginning with $(-11,-4)$: $(-4+11)^2+(-7+4)^2=68$; $49+9\neq68$; the point is not on the circle. Continuing with $(-12,-5)$: $(-4+12)^2+(-7+5)^2=68$; $64+4=68$; the point is on the circle. Continuing with $(5,-8)$: $(-4-5)^2+(-7+8)^2=68$; $81+1\neq68$; the point is not on the circle. Lastly, with $(-5,-12)$: $(-4+5)^2+(-7+12)^2=68$; $1+25\neq68$; the point is not on the circle.

Question No.	Answer	Detailed Explanations
4	A	The equation of a circle is in the form $(x-h)^2+(y-k)^2=r^2$. The general form of the point that is the center of a circle is (h,k), so h=−3, and k=8. This gives us the equation $(x+3)^2+(y-8)^2=r^2$. The question says the circle contains the point (4,3). This means that the radius of the circle is the distance between the points (−3,8) and (4,3). Use the distance formula to find the radius squared. $r=\sqrt{(-3-4)^2+(8-3)^2}=\sqrt{74}$; so $r^2=74$. The equation of the circle is $(x+3)^2+(y-8)^2=74$. Now, substitute the points into the equation for x and y. If the result is a true statement, then the point is on the circle. Beginning with (2,15) : $(-3-2)^2+(8-15)^2=74$; 25+49=74; the point is on the circle. Continuing with (−10,11) : $(-3+10)^2+(8-11)^2=74$; 49+9≠74; the point is not on the circle. Continuing with (−10,12) : $(-3+10)^2+(8-12)^2=74$; 49+16≠74; the point is not on the circle. Lastly, with (−9,15) : $(-3+9)^2+(8-15)^2=74$; 36+49≠74; the point is not on the circle.
5	C	If two lines are parallel, they have the same slope. The slope-intercept form of a linear equation is y=mx+b, where m is the slope and b is the y−intercept. The slope of the given equation $y=\frac{2}{5}x-7$ is $\frac{2}{5}$. Since the questions asks for a line that is parallel to the given line, the answer is a line with the same slope. The question gives a point the new line must pass through, and the given equation gives the slope, so use the point-slope form of the equation to find an equation of the line. The point-slope equation of a line is $y-y_1=m(x-x_1)$, where x and y are the variables in the equation and (x_1,y_1) is the point the line passes through. Thus, the point-slope equation is $y-1=\frac{2}{5}(x-3)$. To convert this equation to the slop-intercept equation, distribute the right side and solve for y. $y-1=\frac{2}{5}(x-3)$; $y-1=\frac{2}{5}x-\frac{6}{5}$; $y=\frac{2}{5}x-\frac{6}{5}+1$; $y=\frac{2}{5}x-\frac{6}{5}+\frac{5}{5}$; $y=\frac{2}{5}x-\frac{1}{5}$
6	B	The correct answer is B due to its size and the significance is that this is when the United States had named itself.
7		Comparing the published copy of the "Declaration of Independence" with the rough draft, Jefferson made many changes throughout. The changes can be seen on the rough draft by his strikethroughs.
8		Most likely the most difficult section to write using the rough draft would be the beginning of the first and second paragraphs. This can be seen by the number of strikethroughs in these areas. Also, he scribbled out several sentences in these places so that they can't be read at all.

Question No.	Answer	Detailed Explanations
9		There is a shaded part in the drawing for the bow because that allows the person to grip the bow before shooting the arrow. The student would need to infer that the shaded part would be drawn for a purpose and that would be where the person would be holding the bow.
10		The drawing reveals one small part of the entire telegraph. It shows what the arm looks like and a description explains where the wires are and the two blank circles indicate the wire which ties the line wire to the insulator. The picture gives a visualization of one part to show how it works.

Week 4

Question No.	Answer	Detailed Explanations
1	C	Point A is at -5 on the number line in the figure, and point B is at 10. Thus, the length of segment AB is 15. To divide the segment into two parts with a ratio of their lengths of 2:1, change the ratio to 2x:1x to allow variation in the location on the number line. Next, set the sum of the two parts equal to 15 and solve for x. $2x+1x=15$; $3x=15$; $x=5$. Now, that you know that $x=5$, find $2x$, which equals 10. Find the value on the number line by adding 10 to the position of point A. $-5+10=5$. The value on the number line that divides segment AB in a ratio of 2:1 is 5.
2	C	Point C is at -30 on the number line in the figure, and point D is at 30. Thus, the length of segment CD is 60. To divide the segment into two parts with a ratio of their lengths of 1:2, change the ratio to 1x:2x to allow variation in the location on the number line. Next, set the sum of the two parts equal to 60 and solve for x. $1x+2x=60$; $3x=60$; $x=20$. Now, that you know that $x=20$, Find the value on the number line by adding 20 to the position of point C; $-30+20=-10$. The value on the number line that divides segment CD in a ratio of 1:2 is -10.
3	C	The coordinates of the vertices are: point $E:(-3,5)$; point $F:(3,5)$; point $G:(3,3)$; and point $H:(-3,3)$. Use the distance formula, $d=\sqrt{(x_2-x_1)^2+(y_2-y_1)^2}$ to find the length of each side. Based on the graph, sides EH and FG have the same length, which is $\sqrt{(-3-(-3))^2+(5-3)^2}=2$. Also, sides EF and HG have the same length, which is $\sqrt{(3-(-3))^2+(5-5)^2}=6$. The sum of the sides is $2+2+6+6=16$. The perimeter is 16.
4	C	Use the formula for volume of a pyramid: $V=\frac{1}{2}\times a\times c\times h$ In this case the length is 15cm, the base is 10 cm in length, and the height is 9 cm. Therefore : $V=\frac{1}{2}\times 15\times 10\times 9=675$ cm³

Question No.	Answer	Detailed Explanations
5	C	Use the formula for volume of a cylinder: $V=\pi r^2 h$. In this case the radius (r) can be found by dividing the diameter in half. If the diameter is 2 meters then the radius will be 1 meter $(\frac{2m}{2}=1m)$. The height (h) is given as 3 meters. This information can now be substituted into the formula. $V=\pi r^2 h$ $V=\pi(1m)^2 3m = 3.14(1m^2)(3m) = 9.42m^3$
6	B	The correct answer is B, logical fallacy.
7	D	The correct answer is D, because these are all different types of logical fallacies.
8	B	The answer is B, because it is about connecting ideas in an illogical way.
9	C	The answer is C because the article offers no real proof to make the generalization that kindness leads to those three things. There might be other factors that lead to this result.
10		Answers can vary, but must include an opposing argument. For example: "Unfortunately, in many schools, negative behaviors such as bullying results in punishment which is thought to reduce this kind of behavior in future." Referencing the zero-tolerance is the counter argument to teaching kindness, and the author debunks its value in the next sentence. This is why it is the counter argument.

Question No.	Answer	Detailed Explanations
1	B	Slicing the cylinder through the middle would result in a circle. 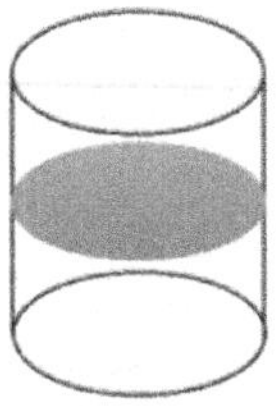
2	B	Cutting the triangular prism in the fashion shown below would result in an equilateral triangle. 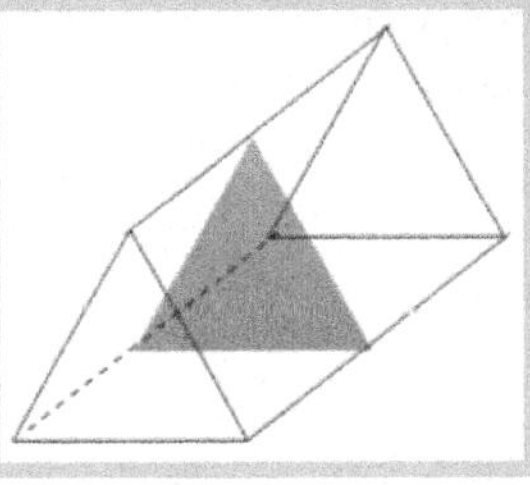
3	B	Remember the formula for perimeter is the sum of the length of the sides. For a square where all sides are equal: P=4s Substituting the value for perimeter 4 ft=4s 1 ft=s If each side measures 1 ft. The area will be: A=s^2 Substituting the value of s: A=(1ft)^2=1ft^2
4	C	In this case find the area of the fence not the yard. So the length will be 20 ft and the height of 3 ft will serve as the other dimension. There will be two sections that have an area of A=20 ft × 3 ft = 60 ft^2 Doubling this: 60 ft^2×2=120 ft^2 The other side of the fence will have an area of A=30 ft × 3 ft = 90 ft^2 Adding these areas results in: A=120 ft^2+90 ft^2=210 ft^2 Now divide this by the amount 1 gallon will cover $\frac{210 \text{ ft}^2}{50 \text{ ft}^2/\text{gallon}}=4.2$ John will need at least 5 gallons.

Question No.	Answer	Detailed Explanations
5	B	The trunk will be in the shape of a cylinder. Therefore $V=\pi r^2 h$ Since the height is known, use the formula for circumference to find the radius $C=\pi d$ 5π feet $= \pi d$ 5feet=d and the radius will be ½ that or 2.5 feet. Use this value in the Volume formula $V=\pi(2.5 \text{ feet})^2(10 \text{ feet}) = 62.5\pi \text{ feet}^3$.
6	D	The correct answer is D, because implicit reading means the reader have to do around to really look at what the writer is trying to say. The ideas are not directly stated.
7	D	The correct answer is D. She lists all of those arguments in the essay.
8	D	The answer is D, because the writer offers no real evidence to support her point.
9	B	The correct answer is B, because she doesn't mention the main points in her conclusion.
10		While there is an argument that 16 and 17-year-olds are too immature to vote, there is also a danger that they might not vote at all.

Question No.	Answer	Detailed Explanations
1	C	Begin by finding the area of the round petri dish. Remember the diameter is twice the radius. $A=\pi r^2$ $A=(3.14)(\frac{10}{2})^2$ $A=(3.14)(5cm)^2$ $A=(3.14)25cm^2=78.5cm^2$ Next use this area to find how many bacteria are present. $78.5cm^2 \times 50 \frac{bacteria}{cm^2}=3,925$
2	B	Begin by finding the volume of the beaker (a cylinder) $V=\pi r^2 h$ $V=(3.14)(4cm)^2(12cm)=(3.14)(16cm^2)(12cm)=602.88 \ cm^3$ Now use the density of water to find the weight of the water. $602.88 \ cm^3 \times \frac{1g}{cm^3}=602.88 \ g$ Since this value represents only the weight of the water in the beaker, and not the beaker itself, the weight of the water and beaker must be added together. Total Weight=Weight of water+Weight of Beaker Total Weight=602.88 g+3g=605.88 g
3	C	Understand that in order to determine how many laps will be made the length of the laps will need to be determined. The length of the laps will be the same as the circumference of the circle. The circumference is $C=\pi d$; $C=\pi(1mile)$; $C=3.14$ miles Now determine how many laps $Laps=\frac{600 \ miles}{3.14 \ miles}=191.08$ Therefore in order to run 600 miles the cars will need to drive a full 192 laps.

Question No.	Answer	Detailed Explanations
4	B	The volume of the cylinder minus the volume of the cone will leave the volume of the cylinder not filled by the cone. The radius of both solids will be 2 cm. Begin by finding the full volume of the cylinder $V=\pi r^2 h$ $V=(3.14)(2cm)^2(10cm)$ $V=125.6 \ cm^3$ Now calculate the volume of the cone. $V=\frac{1}{3}\pi r^2 h=\frac{1}{3}(3.14)(2cm)^2(5cm)=20.93cm^3$ Now subtract. $V_{region}=V_{Cylinder}-V_{cone}=125.6 \ cm^3-20.93cm^3=104.67cm^3$
5	B	Consider each option. Option A -- There are no data points in the 60 column, this statement is false. Option B -- The 90 column has more data points than any other column therefore most students scored 90. This option is correct. Option C-- Since no students scored 60 this would be the score with the fewest student scores. Option C is incorrect. Option D -- There were 2 students scoring 70 and 5 students scoring 80 for a total of 7 students. There are 7 data points in the 90 column as well meaning there were equal number of students scoring either 70 or 80 and students scoring 90. Option D is not correct.
6	D	The correct answer is D. The Declaration of Independence directly showed Great Britain's unjustified actions.
7	A	The correct answer is A, because insulting Great Britain would not have been allowed since there was not freedom of speech during that time.
8		The theme of the Declaration of Independence is that Great Britain's government is unfair and shows favoritism to the British. Each short paragraph describes that the British government is favoring its own citizens and not the colonists living in the United States.
9	D	The document accuses the British government without holding back. This was not allowed or heard of during this time without the countries going to war.
10		The significance of the document is that it accuses the British government without holding back. This was not allowed or heard of during this time without the countries going to war.

Question No.	Answer	Detailed Explanations
1	B	The question asks us to provide the difference between the means of two sets of numbers. We find a mean by adding the numbers and then dividing by the number of numbers in the set. First, we will find the mean of set A: $\frac{2+4+6+8+10+12}{6}=\frac{42}{6}=7$. Next, we will find the mean of set B: $\frac{3+5+7+9+11+13}{6}=\frac{48}{6}=8$. The difference between the two means is 1.
2	B	The question asks us to find the difference in the medians of Set A and Set B. The median of a set of numbers is the center number when the numbers are in order from least to greatest. If there is an even number of numbers in the set, then the median is the average of the two middle numbers in the set. Both sets are displayed as dot plots, which show the quantity and value of each member of the set. The dot plot places them in order from least to greatest. Therefore, we must identify how many numbers are in each set. Both sets have 30 numbers, so we find the 15th and 16th numbers in each set, and use their average as the median. In Set A, the 15th and 16th numbers are both 5. Thus, their average is 5. In Set B, the 15th and 16th numbers are both -5. Thus, their average is -5. The difference between these medians is 10.
3	A	The figure below shows the effect on the mean and median as a result of adding some very small elements to a data set. Since the new elements are very small, they have a significant effect on the mean because their very small values are averaged with the other values in the set. The median is also affected, and moves in the same direction as the mean moves. 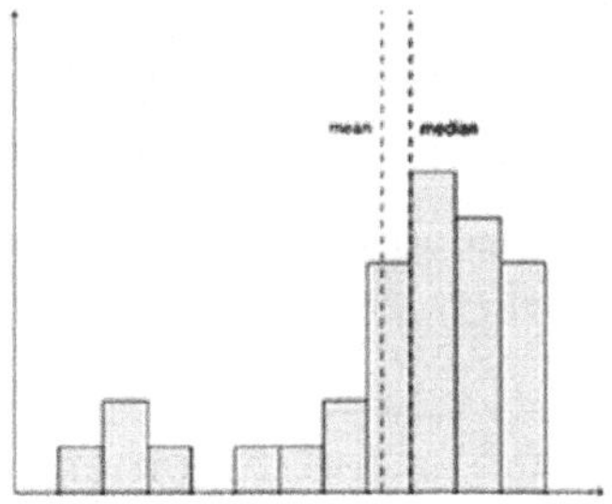

Question No.	Answer	Detailed Explanations
4	B	A uniform distribution is a set of data with all equal values. The histogram below that represents a uniform distribution.

Uniform Distribution

Question No.	Answer	Detailed Explanations
5	D	Since the data is normally distributed the empirical rule can be followed. This rule states that 95% of data lies within 2 standard deviations above and below the mean. Therefore the range that covers 95% will be between $90.12 + (4.3 \times 2) = 98.72$ $90.12 - (4.3 \times 2) = 81.52$
6	B	The answer is B because the phrase contains the verb of the sentence.
7	A	The answer is A because the phrase contains the noun and its modifiers.
8	C	The answer is C because the phrase begins with the preposition "in" and ends with "driveway" as the object of the preposition.
9	B	The answer is B because the sentence is composed of a dependent clause and does not communicate a complete thought.
10	A	The answer is A because the sentence is composed of two independent clauses spliced together with a comma.

Question No.	Answer	Detailed Explanations
1	D	Since the marbles can only be blue or green and 3 marbles are green then 27 must be blue (30-3=27). $$P=\frac{blue}{total}=\frac{27}{30}=0.9=90\%$$ Alternate Explanation: Another way to solve this problem is to find the probability of choosing a green marble and then subtract this from 100%. Since there are only two choices, the probability of both must add to 100%. Therefore… $$P_{green}=\frac{green}{total}=\frac{3}{30}=0.1=10\%$$ $$P_{blue}=100\%-P_{green}$$ $$P_{blue}=100\%-10\%=90\%$$
2	A	There are only 5 students who meet both criteria (6th grade and made an A). Therefore probability can be calculated as: $$P=\frac{6th\ graders\ with\ an\ A}{Total\ students}=\frac{5}{50}=0.10=10\%$$
3	A	Notice that the trend of the graph between the data points forms a line.
4	B	In equations with dependent and independent variables "x" typically serves to represent the independent variable. In Dr. Winthrop's study the temperature is the independent variable. The variable "x" represents temperature.

Question No.	Answer	Detailed Explanations
5	C	Rewrite the given equation in slope intercept form y=mx+b In this case m will need to equal 1 and b will be zero (0) so as to keep the function equivalent to the given equation. y=1x+0 Therefore, The slope (m) is 1 and the y-intercept (b) is 0.
6	C	The answer is C because only proper nouns are capitalized along with A and D.
7	B	The answer is B because the two independent clauses are closely related by the subject, so a semicolon is appropriate to join them, and the transition of "therefore" is also needed. A is a comma splice and C inaccurately uses a colon.
8		The sentence incorrectly uses the possessive "your" at the beginning when it should be the contraction of "you are," or "you're." The error is flipped later on in the sentence right before "contribution." The word should be "your" instead of "you're."
9	C	The answer is C because "weather" should be in place of "whether."
10	D	The answer is D because this is the definition of a comma splice.

Question No.	Answer	Detailed Explanations
1	B	The oval region labeled "H" indicates the space with the hammers. The oval region labeled "S" indicates the space with the screwdrivers. The oval region labeled "L" indicates the space with the pliers. Thus, the region outside of these three oval regions is the empty space in the drawer. The question asks which Venn diagram correctly shades the open space in the drawer after Mark finished putting away his tools. The correct answer is the Venn diagram that shades all of the area in the drawer "U" that is outside of the oval regions that are filled with tools.
2	A	The oval region labeled "S" indicates the days when it snowed. The oval region labeled "R" indicates the days when it rained. The oval region labeled "Y" indicates the days when it was sunny. The oval region labeled "C" indicates the days when it was cloudy. Thus, the region outside of these four oval regions represents the entire month. The question asks which Venn diagram correctly shades the days in which it either snowed or rained. The correct answer, based on the information above, is the Venn diagram that has the oval region labeled "S" and the oval region labeled "R" shaded. The other oval regions represent other types of weather.
3	B	Since a standard deck of playing cards has 52 cards, with 13 hearts, 13 clubs, 13 spades, and 13 diamonds, the probability of picking the first diamond is $\frac{13}{52}$ or $\frac{1}{4}$. Since you did not replace the first card back into the deck, the number of diamonds in the deck changed from 13 to 12 and the number of cards in the deck changed from 52 to 51. The probability of selecting a second diamond is $\frac{12}{51}$. Since this probability is different from the first probability, the two events, picking the first diamond and picking the second diamond are dependent events.
4	B	Since a standard deck of playing cards has 52 cards, with 4 aces, the probability of picking the first ace is $\frac{4}{52}$ or $\frac{1}{13}$. Since you did not replace the first card back into the deck, the number of aces in the deck changed from 4 to 3 and the number of cards in the deck changed from 52 to 51. The probability of selecting a second ace is $\frac{3}{51}$ or $\frac{1}{17}$. Since this probability is different from the first probability, the two events, picking the first ace and picking the second ace are dependent events.

Question No.	Answer	Detailed Explanations
5	D	The table shows the result of the survey. The question asks us to find the probability that a randomly selected student is interested in hiking given the condition that the student is a girl. This is called conditional probability, which is calculated by $$P(\text{Hiking}\mid\text{Girl})=\frac{P(\text{Hiking}\cap\text{Girl})}{P(\text{Girl})}.$$ From the table, $P(\text{Hiking}\cap\text{Girl})=\frac{50}{250}$ and $P(\text{Girl})=\frac{113}{250}$. Therefore, $P(\text{Hiking}\mid\text{Girl})=\frac{\frac{50}{250}}{\frac{113}{250}}=\frac{50}{113}$.
6	C	The answer is C because The Atlantic is listed in the spot where the source title should appear in a Works Cited entry.
7	A	The answer is A because Julie Beck is listed in author position of the Works Cited entry.
8	B	The answer is B because fallible is synonymous with faulty just as credible is synonymous with trustworthy.
9	A	The answer is A because Shane Snow appears in the author position of the Works Cited entry.
10	B	The answer is B because Smartcuts appears in the book title position of the Works Cited entry.

Question No.	Answer	Detailed Explanations
1	C	Note that all of the choices other than C are represented in numbers. However there is no way to numerically represent a color. This is referred to as categorical data.
2	B	$P(\text{male student}) = \dfrac{\text{Number of male students}}{\text{Total number of students}}$ $= \dfrac{40}{40+50} + \dfrac{40}{90}$ $= 0.4444 = 44.44\%$
3	A	Choice A is the only one that talks about snoring and being overweight on all the labels.
4	C	To find the probability that a person prefers peas, you simply need to find P(P).
5	B	To find the probability that a person prefers carrots given that they speed, you will need to find P(C|S).
6	C	Answer is C, because the author's last name and then first name are listed first in the Works Cited entry.
7	D	Answer is D, because The New York Times, the title of a newspaper, is listed in the source title position in the Works Cited entry.
8	C	Answer is C, because the date accessed is not included in the MLA citation of a printed journal article; all other information listed is available in the Works Cited entry.
9	B	Answer is B, because graceful is the opposite of awkward just as excruciating is the opposite of mild.
10	A	Answer is A, because gasoline is stored in a tank just as money is stored in a vault.

Question No.	Answer	Detailed Explanations
1	$\dfrac{1}{4}$	There is only 1 data point (star) that is in circle X out of the 4 data points that are in Y, so the probability is 1 out of 4.
2	B	To find W given X we put the number of data points that are in W and also in X on top, then we put the total number of data points that are in X on bottom. This gives us $\dfrac{2}{12}$ which reduces to $\dfrac{1}{6}$.
3	$\dfrac{5}{24}$	Use the formula for P(B or C), P(B or C)=P(B)+P(C)−P(B and C). $P(B)=\dfrac{2}{24}$ because there are 2 data points in circle B and 24 data points total $P(C)=\dfrac{4}{24}$ because there are 4 data points in circle C and 24 data points total. $P(B \text{ and } C)=\dfrac{1}{24}$ because there is 1 data point in the area where B and C overlap and 24 data points total. Now using the formula we get, $P(B \text{ or } C)=\dfrac{2}{24}+\dfrac{4}{24}-\dfrac{1}{24}=\dfrac{5}{24}$
4	C	Simply count the data points in circles E and F. There are 10 of them out of 24 total data points and by reducing we get $\dfrac{10}{24}=\dfrac{5}{12}$.
5	30 outcomes	To solve this problem, you will need to use the formula for Permutations from above. $$_nP_r=\dfrac{n!}{(n-r)!}=\dfrac{6!}{(6-2)!}=\dfrac{6!}{4!}=\dfrac{6\times5\times4\times3\times2\times1}{4\times3\times2\times1}=6\times5=30$$ The $4\times3\times2\times1$ on top and bottom will cancel each other out so you are left with only $6\times5=30$.
6	B	The answer is B because over means excessively, as in excessively confident, excessively burdened, and excessively joyed.
7	D	The answer is D because highlight is a synonym for accentuate.
8	B	The answer is B because anti- means against.
9	C	The answer is C because aud is the root word meaning sound.
10	D	The answer is D because the suffix -ism means belief.

Week 6

Question No.	Answer	Detailed Explanations
1	C	Use the formula for combinations from above and solve as follows, $$_nC_r = \frac{n!}{(n-r)!r!} = \frac{5!}{(5-5)!5!} = \frac{5!}{5!} = 1$$
2	C	If \$3500 and \$2000 are only 20% of the cost charged, you can calculate the total charge for a broken leg and arm each as $\frac{3500}{.20}$ =\$17,500 and $\frac{2000}{.20}$=\$10,000 respectively. Since there's only a chance these events will occur you can use $E(x)=x_1 p_1 + x_2 p_2 + \cdots + x_i p_i$ and find out that $E(x)=17,500\times.20+10,000\times.40=\$7,500$. Calculating the expected value with insurance gives $E(x)=3,500\times.20+2,000\times.40=\$1,500$. The answer is the difference between the two, \$7,500-\$1,500=\$6,000.
3	A	To calculate the E(x) we simply need to add up all the money amounts and multiply by the likelihood of one of them being randomly drawn since they are all equally likely to be drawn. Calculate as shown below, $E(x)=(90.6+90+83.2+74.3+72.2+69.7+62.1+53.3+53+48.5+48.5)\times\frac{1}{11}=745.4\times\frac{1}{11}=67.76364...$ So this rounds to 68.
4		Your probability of winning is 1 out of 25 or 0.04. We can create a probability distribution with n=30, p=0.04, q=0.96, and a going from 0 to 2 since we are interested in 2 or fewer times of winning. So find P(0), P(1), and P(1) as follows. $$P(X=0)=\frac{30!}{30!0!}\times0.04^0\times0.96^{30}=0.2939...$$ $$P(X=1)=\frac{30!}{29!1!}\times0.04^1\times0.96^{29}=0.3673...$$ $$P(X=2)=\frac{30!}{28!2!}\times0.04^2\times0.96^{28}=0.2219...$$ The probability of all those things happening is the sum of all the probabilities, 0.88. So the probability that you'll win 2 or fewer times buying 30 tickets is 88%.

Question No.	Answer	Detailed Explanations
5	C	Assuming there is only one correct answer, the probability of getting a wrong answer on the first problem is 0.75. The probability of getting a right answer on the second problem is 0.25. Since order matters, if we multiply the probabilities together, that will give us the probability of giving an incorrect answer and then a correct answer. That means our answer is $0.75 \times 0.25 = 0.1875$.
6	A	The answer is A because cheap has more negative associations with it than inexpensive.
7	B	The answer is B because shack carries more of a connotation of a run-down building as opposed to home, which connotes comfort and family.
8	D	The answer is D because "old news," comprised of two contradictory terms, is an oxymoron.
9	B	The answer is B because the sentence is making a comparison between two unlike things, using the word like.
10	D	The answer is D because the sentence gives human characteristics to clouds, calling the rain tears.

Week 6

Question No.	Answer	Detailed Explanations
1	D	Since most months have 30 days we will assume 30 days in a month. We can use $E(x) = x_1 p_1 + x_2 p_2 + \cdots + x_i p_i$ or simply calculate as follows $E(X) = .05 \times 250 \times 30 = \$375$
2	C	Since the probability is 0.7 that you will gain $2 and 0.3 that you will lose $1, twenty bets would give you 14 wins and 6 losses. That translates to $28 gained and $6 lost, which makes a net gain of $22.
3	B	To find the expected value earned for each game you can multiply the probability by the earnings of each card as shown $0.5 \times 0 + 0.25 \times 1 + 0.15 \times 3 + .1 \times 5 = 1.20$ Since it costs $2 to play, the average gain per game is -$0.80. If your cousin plays 10 games, he's expected to earn $12 and pay $20, meaning a net gain of -$8.
4	D	To be a fair decision, any items used in the process should not be different in any way.
5	C	Both A and B are fair ways to randomly make a decision with no bias in any direction.
6	B	The answer is B because formidable means impressive in strength or excellence, which most fits the context.
7	C	The answer is C because flourish means to grow vigorously, which most fits the context.
8	A	The answer is A because jovial means full of or showing high-spirited merriment, which most fits the context.
9	D	The answer is D because unanimous means in complete agreement, which most fits the context.
10	A	The answer is A because voracious means desiring or craving food, which most fits the context.

Question No.	Answer	Detailed Explanations
1		Incorrect, the chances are actually 100% based on this wording.
		If you read the question carefully it says "the chances that the item will be bought by a random person or not be bought by a random person". They are definitely going to be bought...or not bought... so the chances of this scenario are 100%.
2	A	Student must set up a proportion $\frac{0.5 \text{ cup chocochips}}{2 \text{ cups flour}} = \frac{x}{3 \text{ cups flour}}$ and solve for x by cross multiplying and then dividing. So, $x = 0.5 \times \frac{3}{2} = \frac{1.5}{2} = \frac{3}{4}$.
3	D	Student must know the area formula for a rectangle ($A = lw$) and then properly set up an equation that represents the width in terms of the length; answer choice B reverses the order of the phrase "5 less than twice its length" so it is incorrect, choices A and C are different variations of one another making D correct.
4	B	The number of kilometers reached, k, is a function of 3 times the number of minutes, m. Therefore, k=3m.
5	A	Using $y = mx \pm b$ we see the rate (m) is constant at 4 square feet per hour; a point on this line is give as (2, 40), student must solve for b to find it is 48.
6	C	Choice C is the best thesis statement because it clearly states a claim. It is both debatable and supported by evidence.
7	D	The answer is D because an argumentative thesis must be debatable and cannot just state a fact.
8	B	The answer is B because the counterclaim is when the opposing view is considered.
9	B	The answer is B. A conclusion should close with a call-to-action for the audience. This call-to-action should outline what you want the audience to do in response to your argument.
10	B	The answer is B because the other choices use informal language.

Question No.	Answer	Detailed Explanations
1	B	Set up a proportion of guests to the total number of people, $\frac{5}{8}=\frac{x}{576}$. Solve by cross multiplying. $8x = 2880$. Divide both sides by 8. So x=360.
2	C	Student must set up a linear equation such as $\$875 + 20c \leq 35c$ and solve for c by combining like terms and using inverse operations. At the end of the 58th day, James has not broken even. He exceeds the break even point on the 59th day.
3	A	Student must recognize that $55 is the constant in the equation (flat fee). And the additional charges of $40 over 3 gigabyte use. So, the equation is $V = 55 + 40(d - 3)$.
4	B	Student must apply inverse operations by multiplying both sides of the equation by 2 and then divide by h to isolate b. Therefore, $b = \frac{2A}{h}$
5	C	Student must apply the inverse operation of division to isolate h; dividing both sides of the equation by πr^2 will isolate h. Therefore $h=V- \pi r^2$
6	Thesis	The most important sentence(s) in your argument is Thesis.
7		Answers can vary, but must include some form of 1) pre-writing/brainstorming/mental mapping, 2) drafting 3) revising 4) proofreading 5) publishing
8	D	The correct answer is D because all of those parts are in a traditional paragraph.
9	C	The correct answer is C because the three points addressed in the thesis correspond to the three supporting body paragraphs that follow the introduction.
10	C	The correct answer is C because the thesis should be the last line of your introduction.

Week 7 Summer Practice

Day 1

Question No.	Answer	Detailed Explanations
1	B	Student must solve this equation with variables on both sides for x by applying inverse operations and grouping like terms on either side of the equal sign; alternatively a student might test each of the answer choices in the equation using substitution and find which one balances the equation. $2x-4=3x-11$ $2x-4-2x=3x-11-2x$ $-4=x-11$ $-4+11=x-11+11$ $x=7$
2	D	Student must demonstrate knowledge of the multiplicative inverse which is the reciprocal of any real number; student must understand that the product of a number and its multiplicative inverse is one; this is commonly used in solving equations that involve fractions.
3	B	Student must first solve the one-step equation by applying the inverse operations of addition and then division resulting in the inequality $x<4$; answer choice B shows the positive integers less than 4. $2x-3+3<5+3$ $2x<8$ $x<4$
4	B	There is more than one way to get the correct table of values but the simplest way is to type the equation into a graphing calculator and look at the table of values. You could also check each x-value individually by plugging it into the equation for x and finding y. This could be quite time consuming since many of the table have different x-values.
5	C	This problem can be easily solved by factoring out the greatest common factor 4x and then solving for each factor as shown: $4x^2-8x=0$ $4x(x-2)=0$ $4x=0$ and $x-2=0$ $x=0$ and $x=2$

Question No.	Answer	Detailed Explanations
6	B	The answer is B because A is narrative, C is an argument, and D is a poem.
7	C	The answer is C because the author presents an effect and uncovers the underlying cause of the effect.
8	A	The answer is A because the author presents a problem and then a clear solution to fix it.
9	D	The answer is D because the author gives the reader instructions on how to complete a task.
10	B	The answer is B because the author is comparing both the similarities between desktop and laptop computers and the differences.

Day 2

Question No.	Answer	Detailed Explanations
1	C	The elimination method (also known as the addition method) allows you to add the two equations together thereby eliminating one of the variables so that you can solve for the one that is left. Then use the one that you know to help you find the other one. $5x+2y=0$; $3x-2y=-16$; $8x=-16$; $x=-2$ Now use that x-value and plug back into one of the original two equations to find y. $5x+2y=0$; $5(-2)+2y=0$; $-10+2y=0$; $2y=10$; $y=5$ So then the final answer is $(-2, 5)$ or choice C.
2	D	There are several way to solve this system of equations. If you solve the bottom equation for y, you get $y=4-x$. Then you can use substitution by plugging in $4-x$ in place of y in the first equation like so and then solve for x. $-2x=y-1$; $-2x=(4-x)-1$; $-2x=4-x-1$; $-2x=3-x$; $-x=3$ $x=-3$ Now use $x=-3$ and plug into one of the original equations to find y. $y+x=4$; $y+-3=4$; $y=7$ So the intersection point and final solution is $(-3, 7)$.
3	C	You can set up a system of equations to solve this problem. We will let k equal the number of miles Kim ran and a equal the number of miles Alex ran. The two equations would be as follows $k - a = 4$; $k + a = 20$ Using the elimination (addition) method, you could eliminate the a's and solve for k. $k - a = 4$; $k + a = 20$; $2k = 24$; $k = 12$ Now use the fact that $k=12$ and plug that value back into one of the original equations to find a. $k + a = 20$; $12 + a = 20$; $a = 8$ So the intersection point and final answer is $(12, 8)$.

Question No.	Answer	Detailed Explanations
4	A	You can set up a system of equations to solve this problem. We will let p equal the cost of one can of paint and r equal the cost of one roller. The two equations would be set up as follows $2p + r = 62$; $5p + 2r = 151$ Now you can multiply the top equation by -2 to create r's that are opposites and would eliminate. $-2(2p + r = 62)$; $-4p - 2r = -124$ Now add this changed equation to the original bottom equation $-4p - 2r = -124$; $5p + 2r = 151$; $p = 27$ Now use the fact p=27 and one of the original equations to find r. $2p + r = 62$; $2(27) + r = 62$; $54 + r = 62$; $r = 8$ So the intersection point and final answer is (27, 8).
5	B	To solve this system of equations, start by substituting $\frac{1}{4}x-8$ or 0.25x - 8 in for y in the first equation like so, $y = x^2 + 5x + 9$; $0.25x - 8 = x^2 + 5x + 9$; $0 = x^2 + 4.75x + 17$ Now use the discriminant b^2-4ac to determine how many solutions this system would have. $4.75^2 - 4(1)(17) = -45.4375$ Since the discriminant is a negative number, there would be no real solutions and the two graphs would not intersect.
6	B	The answer is B because the protagonist is the main character of the story.
7	D	The answer is D because the author uses third person pronouns and focuses only on the thoughts and emotions of one character.
8	A	The answer is A because this is the part of the plot diagram in which the setting and characters of a story are introduced.
9	C	The answer is C because a flashback takes the reader to a scene earlier in or before the story.

10

	First Person	Second Person	Third Person
I, We	◯		
Me, Us	◯		
He, she, it			◯
You, yours		◯	

I, we, me and us are first-person pronouns that refer to himself/herself. He, she and it are third person pronouns and you, yours are second person pronouns.

Day 3

Question No.	Answer	Detailed Explanations
1		Karen wasn't wrong until the 3rd step of her solution. When you multiply -9(-6) you should get +54 and Karen put -54. $y=4-9x$ $y=4-9(-6)$ $y=4+54$ $y=58$

For Question 2:

	13	-21	-12
$y=8x-5$		○	
$y=-3x+7$	○		
$y=-x-14$			○

Simply plug in a -2 for each of the x values in the three equations and solve for y.
$y=8x-5$
$y=8(-2)-5$
$y=-16-5$
$y=-21$ and continue on in this manner for all three equations.

Question No.	Answer	Detailed Explanations
3	A	You can see by the graph that the two functions never intersect at all.

		(0, 2)	(1, 7)	(5, 1)
Question No.	Answer		Detailed Explanations	
4	A & B	$y_1 > 3x - 2$	✓	✓	
		$y_2 \leq -x^2 + 5$	✓		
		$y_4 \leq \frac{-1}{4}(x-5)^2 + 4$			✓
		$y_3 > -x + 4$		✓	✓

You will need to check to see if each point makes the inequality true as shown below.
For the point $(0,2)$: $2 > 3(0) - 2 \Rightarrow 2 > -2$, yes
For the point $(1,7)$: $7 > 3(1) - 2 \Rightarrow 7 > 1$, yes
For the point $(5,1)$: $1 > 3(5) - 2 \Rightarrow 1 > 13$, no
For $y_2 \leq -x^2 + 5$
For the point $(0,2)$: $2 \leq -0^2 + 5 \Rightarrow 2 \leq 5$, yes
For the point $(1,7)$: $7 \leq -1^2 + 5 \Rightarrow 7 \leq 4$, no
For the point $(5,1)$: $1 \leq -5^2 + 5 \Rightarrow 1 \leq -20$, no
For $y_3 > -x + 4$
For the point $(0,2)$: $2 > -(0) + 4 \Rightarrow 2 > 4$, no
For the point $(1,7)$: $7 > -(1) + 4 \Rightarrow 7 > 3$, yes
For the point $(5,1)$: $1 > -5 + 4 \Rightarrow 1 > -1$, yes
For $y_4 \leq \frac{-1}{2}(x-5)^2 + 4$

For the point $(0,2)$: $2 \leq \frac{-1}{2}(0-5)^2 + 4 \Rightarrow 2 \leq -8.5$, no

For the point $(1,7)$: $7 \leq \frac{-1}{2}(1-5)^2 + 4 \Rightarrow 7 \leq -4$, no

For the point $(5,1)$: $1 \leq \frac{-1}{2}(5-5)^2 + 4 \Rightarrow 1 \leq 4$, yes
Anytime you get a true statement (yes), then the point will be a solution to the inequality.

5	D	We know that there are 12 months in a year, so the most soccer balls your team would need would be 25 or 12 + 13. It is important to remember the meaning of an inequality. Just because you might need as many as 25 balls this year does not make the other answer choices incorrect. All of the answers given fit the inequality, "y is less than or equal to 25". Therefore we must say that all of the answers are acceptable answers to this question.
6	C	The answer is C because the author is providing information about bats.
7	A	The answer is A because the author uses the key word "should" and wants the reader to do something, using an argument for why they should.

Question No.	Answer	Detailed Explanations
8	C	The answer is C because A is missing a comma before "but" and uses "wood" instead of "would," B is missing an ending period, and D is missing the beginning capitalization and the comma before "but." C is written correctly.
9	A	The answer is A, or True, because a narrative is concerned with telling a story and mostly to entertain, as opposed to an argument.
10	D	The answer is D because A, B, and C should all be capitalized.

Week 7

Question No.	Answer	Detailed Explanations
1	$4x^2$	To solve this problem, you will need to plug function g into function f as shown below. $f(g(x)) = x^2$ This is the most simplified answer: $f(2x) = (2x)^2$ $f(g(x)) = 4x^2$
2	C	When solving for the final velocity of an object with a known initial velocity and a constant acceleration over time, we know that the formula is initial position plus velocity times time. So answer choice C it correct.
3	146	Here, a_1 = first term = 1; d = common difference = 6 - 1 = 5; n = number of terms = 30; Now, substitute the value of a_1, d and n in the formula, $a_n = a_1 + (n - 1)d$ $a_{30} = 1 + (30 - 1)(5)$ $a_{30} = 1 + (29)(5)$ $a_{30} = 1 + 145$ $a_{30} = 146$ Alternate Explanation- The pattern here is to add 5 to the previous term, with the exception of term 1 which is given. You will need to extend this pattern to the 30th term. The explicit and recursive equation would be $f(n+1) = f(n) + 5$, but since we need the 29th term to find the 30th term, we will have to extend the pattern all the way out as shown below starting with term 5 21, 26, 31, 36, 41, 46, 51, 56, 61, 66, 71, 76, 81, 86, 91, 96, 101, 106, 111, 116, 121, 126, 131, 136, 141, 146

Question No.	Answer	Detailed Explanations
4	B	The first year you will get $10, so you have to pick the function that has a function value of 10 when you plug in a 1 for n. This eliminates choices A and D as these both give a value of $12.50 when you plug in a 1 for n. To find the correct answer between B and C, we will need to plug in a 2. The second year you should get $12.50. The only one that gives you $12.50 when you plug in a 2 for n is answer choice B. Alternate Explanation - Here, a_1 = first term = 10; d = common difference = 2.50; n = number of terms; Now, substitute the value of a_1, d and n in the formula, $a_n = a_1 + (n - 1)d$ $a_n = 10 + (n - 1)(2.50)$ $a_n = 10 + 2.50n - 2.50$ $a_n = 7.50 + 2.50n$ Therefore, correct answer is option B.
5	A	The value added to the function causes a vertical shift in the graph. Since 12 is 5 units larger than 7, the graph of g(x) is obtained by shifting f(x) 5 units up.
6	A	The answer is A because it is the only sentence that has only one independent clause, though someone might be thrown off by the compound verb.
7	B	The answer is B because it is the only sentence with two independent clauses joined by a coordinating conjunction.
8	B	The answer is B because A has an introductory dependent clause, and so does C, though C is missing the required comma after an introductory clause or phrase. D is also missing the required comma even though it does begin with an introductory phrase.
9	C	The answer is C because editing is about correcting errors and C is about revising writing to make it better, not correcting a mistake.
10	C	The answer is C because revising is about improving writing, as in A, B, and D, but C is about correcting errors.

Question No.	Answer	Detailed Explanations
1	C	To find the inverse, we first need to change the function notation f(x) to a y and then switch the x and y in the equation and solve for y as shown below. $$y=\sqrt{3x-2}$$ $$x=\sqrt{3y-2}$$ $$x+2=\sqrt{3y}$$ $$(x+2)^2=3y$$ $$\frac{(x+2)^2}{3}=y$$
2	A	If we switch the x and y in the function, we end up with $x=3^y$. The only way to solve for y is to take the log. We know that $a=b^c$ means $\log_b a=c$. Using that same rule for this inverse function we end up with answer choice A.
3	C	Answer is C. Each of the side lengths of ΔDEF are twice as long as the ΔABC, therefore the scale factor is 2.
4	B	Answer is B. The definition of scale factor is given in the question
5	C	Answer is C. The original line segment has a length of 2 units. If the scale factor of the new image is s, the new line segment must have a length of 4.
6		The correct order of the writing process is: Planning, Drafting, Revising, Proofreading, Publishing.
7		Answers can vary but should include the idea that the process allows you to edit the information you are sharing so you can choose your structure and organization correctly and proofread. This allows your ideas to be read more clearly.
8		Students might opt to write drafting here, but the correct answer is revising. This is the point in which they take the raw thoughts from the draft and shape it into something stronger.
9	False	The answer is , false, because the stage you focus in on punctuation and capitalization is proofreading.
10	C	The correct answer is C because all of the punctuation is correct in that one. In the first one, it should end in a ?, which eliminates a and b, and in the last one, the writer has created a fragment by putting a period after voting.

Week 8 Summer Practice

Day 1

Question No.	Answer	Detailed Explanations
1	C	Answer is C. Similar shapes have congruent angle measures and proportional sides.
2	B	Answer is B. In order for triangles to be similar, the angle measures need to be the same. Since 56 degrees is not congruent to the measure of angle B it is not similar.
3	C	Answer is C. It is indicating two of the corresponding angles in two triangles are congruent.
4	C	Answer is C. Since AB and ST are parallel, the corresponding angles can be found to be congruent and therefore the triangles are similar.
5	B	Based on the angle markings in the figure, $\angle X$ and $\angle Y$ have the same measure. These two angles are the base angles of $\triangle XYZ$, and therefore they are congruent angles. If the two base angles of a triangle are congruent, then the sides of the triangle opposite those angles are congruent. Thus, side XZ is congruent to side YZ, and congruent sides have the same length. Therefore, if the length of side XZ is 9, then the length of side YZ, is also 9.
6	D	The answer is D because it is the only choice that has a correctly formatted citation at the end of the sentence with the punctuation coming after the citation. The citation also includes the author's last name and the page number from which the quote was found.
7	A	The answer is A because MLA stands for Modern Language Association.
8	D	The answer is D because each of the programs is a valid word processing software to use.
9	C	The answer is C because paraphrasing is when you put someone else's words into your own words.
10	True	The answer is True, because any time you use information from a source, you must give credit to that source.

Question No.	Answer	Detailed Explanations
1	C	The hash marks on the sides of the triangle show congruence with the corresponding side of the triangles. The two congruent triangles can be identified by the pattern of the hash marks and the right angle in the triangles. Notice that on one side of the right angle, sides EF and HI are marked as congruent, and on the other side of the right angle, sides FD and IG are marked congruent. Then, sides ED and HG are marked congruent. Therefore, the congruency statement is $\triangle DEF \cong \triangle GHI$.
2	B	The hash marks on the sides of the triangle show congruence with the corresponding side of the triangles. The two congruent triangles can be identified by the pattern of the hash marks and the appearance of the angles in the triangles. In the figure, $\triangle JKL$ is clearly a different shape than either of the other triangles. Notice that in the other triangles, sides DE and HG are marked as congruent, sides EF and HI are marked congruent. Then, sides DF and GI are marked congruent. Therefore, the congruency statement is $\triangle DEF \cong \triangle GHI$.
3	B	The two triangles are similar using the AA similarity theorem because two corresponding angles in one triangle are congruent to two corresponding angles in the other triangle. When writing a triangle similarity statement, pay particular attention to which angles correspond to each other. The angles with the same measure correspond. Therefore, $\angle H$ corresponds to $\angle M$, $\angle I$, which has a measure of 95°, corresponds to $\angle K$, and then, by default, $\angle J$ corresponds to $\angle L$, which has a measure of 30°. Then, write the similarity statement lining up the corresponding angles in the same order in each triangle: $\triangle HIJ \sim \triangle MKL$
4	B	The question states that $\triangle ABC \sim \triangle XZY$. This means that corresponding sides in one triangle form the same ratio as corresponding sides in the other triangle. You are asked to find the length of side YZ. Set up a proportion to find the length of the side. $\frac{5}{20} = \frac{7}{YZ}$. Cross multiply and find that $YZ = \frac{7 \times 20}{5} = 28$. The length of side YZ=28 cm.
5	A	In a right triangle, the sine of one of the acute angles is the same as the cosine of the other acute angle. The trigonometric ratios in the triangle are $\cos\theta = \frac{Adjacent}{Hypotenuse}$ and $\sin\theta = \frac{Opposite}{Hypotenuse}$. Therefore, sin $R = \frac{16}{20} = \frac{4}{5}$, cos $R = \frac{12}{20} = \frac{3}{5}$, sin $Q = \frac{12}{20} = \frac{3}{5}$, and cos $Q = \frac{16}{20} = \frac{4}{5}$.

Question No.	Answer	Detailed Explanations
6	D	The answer is D because the inquiry question helps you to narrow down your research, so that you're not looking for any information about your subject, but specific information about the subject.
7	D	The answer is D because a works cited page citation requires A, B, and C.
8	D	The answer is D because A,B, and C would all require doing research in order to write.
9	True	The answer is True, because the question doesn't have you researching a specific environment.
10		A credible source will have a recognizable author. If the information is found online, more credible sources are found on websites that end with .edu or .gov. The date of publication is also important in order to help determine relevance of the information.

Question No.	Answer	Detailed Explanations
1	C	The question states that a cable is connected to a pole 12 meters above the ground and extended to 35 meters away from the base of the pole. You are asked to find the angle formed by the cable and the ground. Call that angle θ. The pole forms a right angle with the ground, so use right triangle trigonometry to find the angle. The two values given are the opposite side and the adjacent side of angle θ. Thus, the trigonometric ratio to use is the tangent ratio. Using this ratio, you have $\tan\theta = \frac{opposite}{adjacent}$. Write an equation using $\tan\theta$ and solve for θ. $\tan\theta = \frac{12}{35}$; $\theta = \tan^{-1}(\frac{12}{35})$. Use a calculator to find that $\theta = 18.9°$.
2	B	Use the Pythagorean Theorem to find the length of the hypotenuse. The theorem uses the formula $a^2+b^2=c^2$ where a and b are the sides of the triangle and c is the hypotenuse. Substituting the values given in the question, the formula changes to $12^2+35^2=c^2$. The variable c is the length of the hypotenuse. Solving or c:$144+1225=c^2$; $c^2=1369$; $c=37$. The length of the hypotenuse is 37.
3	B	The weekly interval is not linear, because it is not growing at the same rate every week. This implies exponential growth. It is important to note that in a workout scenario this would not continue to grow exponentially indefinitely.
4	B	Since Jamie will be remaining at a constant speed, this would be a horizontal line on the graph making it linear.
5	B	The only one of these that would be 20 for 0 years is choice B. This function also works for x = 1 and x = 2.
6	B	The answer is B because an autobiography is a first-hand account of someone's life.
7	D	The answer is D because A, B, and C are all first-hand accounts while a biography is a secondary account.
8		Primary sources are first-hand accounts of information or events while secondary sources are second-hand accounts.
9	A	The answer is A because in-text citations are a way for the author to properly give credit to a source whose information the author used.
10		Answers may vary. An important reason to gather information from multiple sources is to help eliminate any potential bias in the author. An author with an open mind and a willingness to look at different perspectives is often able to find insights that biased authors will typically be blind to.

Question No.	Answer	Detailed Explanations
1	B	Linear functions will have an x to the first power. There will be no variables in the exponent and no x to the second power.
2	B	The number of widgets in the warehouse is calculated by subtracting the number of widgets sold from the beginning inventory of widgets. The problem states that the warehouse has 950 widgets when the business opened. The number of widgets sold is calculated by multiplying the sales per week by the number of weeks. The problem also states that the business expects to sell 45 widgets every week and the variable w represents the number of weeks. Therefore, the function that represents the number of widgets in the warehouse after w weeks is I(w)=950−45w.
3	B	Net profit is the result of subtracting the expenses from the total revenue. The problem says the Thespian troupe spent $855 to produce the play, so expenses total $855. The revenue is the price of each ticket times the number of tickets sold, so the total revenue is $20t. Therefore, the function that represents net profit is P(t)=$20t−$855.
4	D	The circumference of a circle is calculated by multiplying the 2π r, where 2π is the complete circle in radians. Thus The length of an arc is calculated a portion of the circumference, so the formula for the length of an arc is the portion of the circle (called $\angle\theta$), times the radius, as long as $\angle\theta$ is measured in radians. The given angle is measured in degrees, so convert the measure to radians by $36°\times\dfrac{\pi}{180°}=\dfrac{\pi}{5}$ radians. Thus, the length of the arc subtended by an angle of $\dfrac{\pi}{5}$ radians on a circle with a radius of 15 is $\dfrac{\pi}{5}\times15=3\pi$ inches.
5	C	The circumference of a circle is calculated by multiplying the 2π r, where 2π is the complete circle in radians. Thus, the length of an arc is calculated a portion of the circumference, so the formula for the length of an arc is the portion of the circle (called $\angle\theta$), times the radius, as long as $\angle\theta$ is measured in radians. Thus, the length of the arc subtended by an angle of $\dfrac{2\pi}{7}$ radians on a circle with a radius of 14 centimeters is $\dfrac{2\pi}{7}\times14=4\pi$ centimeters.

Question No.	Answer	Detailed Explanations
6	B	The answer is B because the author is cautioning the audience against allowing teenagers to overuse smartphones.
7		Answers will vary. Sample: Though smartphones are widely used and provide a number of conveniences, teenagers especially are showing signs of overuse of the technology. Students are being distracted both at school and at home by their smartphones. Parents should talk to their kids about responsible smartphone use.
8	D	The answer is D because the author's focus throughout the passage is how smartphones are distracting even though they are also useful.
9	D	The answer is D because Paragraph 4 considers the opposition to the author's view.
10		The author uses research from studies on sleep deprivation due to smartphone use and other anecdotal evidence of smartphones distracting teenagers from their education and home lives.

Question No.	Answer	Detailed Explanations
1	B	

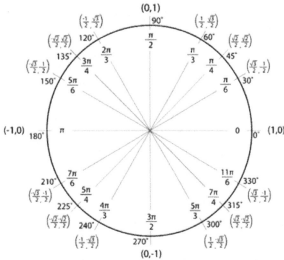

The trigonometric ratio of sine is the ratio of the length of the opposite side divided by the length of the hypotenuse. The length of the opposite side is the y−value in a point on the unit circle. The hypotenuse is the radius of the unit circle, so the hypotenuse is 1. Thus, the value of the sine ratio of any angle in the unit circle is the y−value of the point on the unit circle that corresponds to that angle. The trigonometric ratio of cosine is the ratio of the length of the adjacent side divided by the length of the hypotenuse. The length of the adjacent side is the x−value in a point on the unit circle. The hypotenuse is the radius of the unit circle, so the hypotenuse is 1. Thus, the value of the cosine ratio of any angle in the unit circle is the x−value of the point on the unit circle that corresponds to that angle. In this question, $\sin(\frac{\pi}{3})=\frac{\sqrt{3}}{2}$. This ratio is taken from the point $(\frac{1}{2}, \frac{\sqrt{3}}{2})$ that corresponds to the angle with a measure of $\frac{\pi}{3}$ radians. Thus, using the information above, the value of $\cos(\frac{\pi}{3})$ is the same as the x−value in the point $(\frac{1}{2}, \frac{\sqrt{3}}{2})$. Therefore, the value of $\cos(\frac{\pi}{3})=\frac{1}{2}$.

Question No.	Answer	Detailed Explanations
2	C	

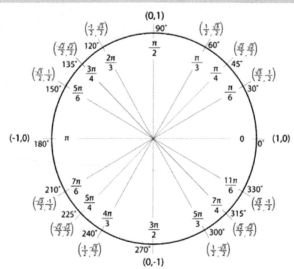

The trigonometric ratio of sine is the ratio of the length of the opposite side divided by the length of the hypotenuse. The length of the opposite side is the y−value in a point on the unit circle. The hypotenuse is the radius of the unit circle, so the hypotenuse is 1. Thus, the value of the sine ratio of any angle in the unit circle is the y−value of the point on the unit circle that corresponds to that angle. The trigonometric ratio of cosine is the ratio of the adjacent side divided by the hypotenuse. The length of the adjacent side is the x−value in a point on the unit circle. The hypotenuse is the radius of the unit circle, so the hypotenuse is 1. Thus, the value of the cosine ratio of any angle in the unit circle is the x−value of the point on the unit circle that corresponds to that angle. In this question, $\cos(\frac{\pi}{3})=\frac{1}{2}$ .This ratio is taken from the point $(\frac{1}{2},\frac{\sqrt{3}}{2})$ that corresponds to the angle with a measure of $\frac{\pi}{3}$ radians. Thus, using the information above, the value of $\sin(\frac{\pi}{3})$ is the same as the x−value in the point $(\frac{1}{2},\frac{\sqrt{3}}{2})$. Therefore, the value of $\sin(\frac{\sqrt{3}}{2})=\frac{1}{2}$.

Question No.	Answer	Detailed Explanations
3	C	An angle with radian measure of $\frac{\pi}{6}$ is the same as a $30°$ angle. This angle can be represented in the first quadrant with a $30-60-90$ triangle with the long leg on the x−axis and the $30°$ angle at the origin, as shown below.

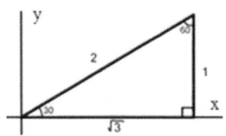

The value of the cosine of an angle is the ratio of the length of the adjacent side to the length of the hypotenuse. In the 30-60-90 triangle, the length of the side that is adjacent to the $30°$ angle is $\sqrt{3}$, which is positive because the x−value in the first quadrant is positive, and the length of the hypotenuse is 2. Therefore, $\cos\frac{\pi}{6}$ $=\frac{\sqrt{3}}{2}$.

| 4 | B | An angle with a measure of $135°$ can be represented in the second quadrant with a 45-45-90 triangle with the one leg on the x−axis and the $45°$ angle at the origin, as shown below. |

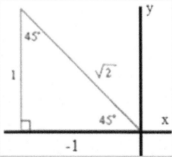

The value of the sine of an angle is the ratio of the length of the opposite side to the length of the hypotenuse. In the 45-45-90 triangle, the length of the side that is opposite to the $45°$ angle is 1, which is positive because the y−value is positive in the second quadrant, and the length of the hypotenuse is $\sqrt{2}$. Therefore, sin $135°=\frac{1}{\sqrt{2}}=\frac{\sqrt{2}}{2}$.

Question No.	Answer	Detailed Explanations
5	C	

The trigonometric ratio of sine is the ratio of the length of the opposite side divided by the length of the hypotenuse. The length of the opposite side is the y–value in a point on the unit circle. The hypotenuse is the radius of the unit circle, so the hypotenuse is 1. Thus, the value of the sine ratio of any angle in the unit circle is the y–value of the point on the unit circle that corresponds to that angle. The trigonometric ratio of cosine is the ratio of the length of the adjacent side divided by the length of the hypotenuse. The length of the adjacent side is the x–value in a point on the unit circle. The hypotenuse is the radius of the unit circle, so the hypotenuse is 1. Thus, the value of the cosine ratio of any angle in the unit circle is the x–value of the point on the unit circle that corresponds to that angle. Select the point on the unit circle that corresponds to the angle $\frac{\pi}{3}$, which is $(\frac{1}{2}, \frac{\sqrt{3}}{2})$. This means $\sin\frac{\pi}{3}=\frac{\sqrt{3}}{2}$ and $\cos\frac{\pi}{3}=\frac{1}{2}$. Then, add $\frac{\pi}{2}$ to $\frac{\pi}{3}$ resulting in $\frac{5\pi}{6}$. The point on the unit circle that corresponds to the angle $\frac{5\pi}{6}$ is $(-\frac{\sqrt{3}}{2}, \frac{1}{2})$. This means $\sin\frac{5\pi}{6}=\frac{1}{2}$ and $\cos\frac{5\pi}{6}=-\frac{\sqrt{3}}{2}$. Therefore, $\sin(\frac{\pi}{2}+\frac{\pi}{3})=\frac{1}{2}$, which is equal to $\cos\frac{\pi}{3}$.

Question No.	Answer	Detailed Explanations
6	D	The answer is D because the passage both persuades the reader of the benefits of a later school start time and supports that claim with general factual information.
7	D	The answer is D because it captures the main idea of the entire passage.
8	B	The answer is B because the color and sound words appeal to the reader's sense of sight and sound.
9	C	The answer is C because it summarizes the most important idea conveyed throughout the entire passage.
10	B	The answer is B because "predetermined" means fixed or prearranged.

Question No.	Answer	Detailed Explanations
1	D	

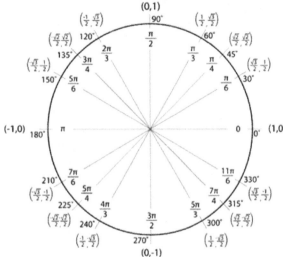

The trigonometric ratio of sine is the ratio of the length of the opposite side divided by the length of the hypotenuse. The length of the opposite side is the y−value in a point on the unit circle. The hypotenuse is the radius of the unit circle, so the hypotenuse is 1. Thus, the value of the sine ratio of any angle in the unit circle is the y−value of the point on the unit circle that corresponds to that angle. In this question, $\sin(\frac{5\pi}{6})=\frac{1}{2}$. This ratio is taken from the point $(-\frac{\sqrt{3}}{2}, \frac{1}{2})$ that corresponds to the angle with a measure of $\frac{5\pi}{6}$ radians. Then, using the information in the unit circle above, the value of $\sin(\frac{5\pi}{6})$ is the same as the y−value in the point $(-\frac{\sqrt{3}}{2}, \frac{1}{2})$. Therefore, the value of $\sin(\frac{5\pi}{6})=\frac{1}{2}$, which is the same as $\sin(\frac{\pi}{6})$ because of the symmetry in the unit circle.

Question No.	Answer	Detailed Explanations
2	C	

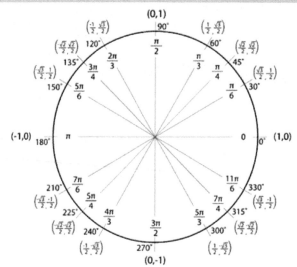

The trigonometric ratio of cosine is the ratio of the length of the adjacent side divided by the length of the hypotenuse. The length of the adjacent side is the x−value in a point on the unit circle. The hypotenuse is the radius of the unit circle, so the hypotenuse is 1. Thus, the value of the cosine ratio of any angle in the unit circle is the x−value of the point on the unit circle that corresponds to that angle. In this question, $\cos(\frac{11\pi}{6})=\frac{\sqrt{3}}{2}$. This ratio is taken from the point $(\frac{\sqrt{3}}{2}, -\frac{1}{2})$ that corresponds to the angle with a measure of $\frac{11\pi}{6}$ radians, which can also be called an angle with a measure of $-\frac{\pi}{6}$ radians. Then, using the information in the unit circle above, the value of $\cos(\frac{\pi}{6})$ is the same as the x−value in the point $(\frac{\sqrt{3}}{2},\frac{1}{2})$. Therefore, the value of $\cos(\frac{\pi}{6})=\frac{\sqrt{3}}{2}$, which is the same as $\cos(-\frac{\pi}{6})$ because the cosine function is an even function.

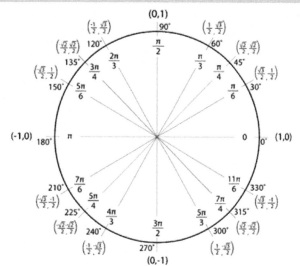

The trigonometric ratio of tangent is the length of the opposite side divided by the length of the adjacent side. The length of the opposite side is the y−value in a point on the unit circle and the length of the adjacent side is the x−value in a point on the unit circle. The hypotenuse is the radius of the unit circle, so the hypotenuse is 1. Thus, the value of the tangent ratio of any angle in the unit circle is the ratio $\frac{y}{x}$ from the point on the unit circle that corresponds to that angle. In this question, $\tan(\frac{\pi}{3})=\sqrt{3}$. This ratio is taken from the point $(\frac{1}{2}, \frac{\sqrt{3}}{2})$ that corresponds to the angle with a measure of $\frac{\pi}{3}$ radians. Then, using the information in the unit circle above, the value of $\tan(-\frac{\pi}{3})$ is the same as the ratio $\frac{y}{x}$ from the point $(\frac{1}{2}, \frac{\sqrt{3}}{2})$. Therefore, the value of $\tan(-\frac{\pi}{3})=\frac{-\frac{\sqrt{3}}{2}}{\frac{1}{2}}=-\sqrt{3}$, which is the opposite of $\tan(\frac{\pi}{3})$ because the tangent function is an odd function with symmetry about the origin.

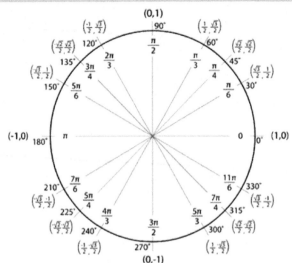

The trigonometric ratio of sine is the ratio of the length of the opposite side divided by the length of the hypotenuse. The length of the opposite side is the y−value in a point on the unit circle. The hypotenuse is the radius of the unit circle, so the hypotenuse is 1. Thus, the value of the sine ratio of any angle in the unit circle is the y−value of the point on the unit circle that corresponds to that angle. In this question, $\sin 60° = \frac{\sqrt{3}}{2}$. This ratio is taken from the point $(\frac{1}{2}, \frac{\sqrt{3}}{2})$ that corresponds to the angle with a measure of 60°. Then, using the information in the unit circle above, the value of $\sin 120°$ is the same as the y−value in the point $(-\frac{1}{2}, \frac{\sqrt{3}}{2})$. Therefore, the value of $\sin 120° = \frac{\sqrt{3}}{2}$, which is the same as $\sin 60°$ because of the symmetry in the unit circle.

Question No.	Answer	Detailed Explanations
5	C	

The trigonometric ratio of cosine is the ratio of the length of the adjacent side divided by the length of the hypotenuse. The length of the adjacent side is the x−value in a point on the unit circle. The hypotenuse is the radius of the unit circle, so the hypotenuse is 1. Thus, the value of the cosine ratio of any angle in the unit circle is the x−value of the point on the unit circle that corresponds to that angle. In this question, $\cos 45° = \frac{\sqrt{2}}{2}$. This ratio is taken from the point $(\frac{\sqrt{2}}{2}, \frac{\sqrt{2}}{2})$ that corresponds to the angle with a measure of 45°. Then, using the information in the unit circle above, the value of $\cos 135°$ is the same as the x−value in the point $(-\frac{\sqrt{2}}{2}, \frac{\sqrt{2}}{2})$. Therefore, the value of $\cos 135° = -\frac{\sqrt{2}}{2}$, which is the opposite of $\cos 45°$ because of symmetry in the unit circle and the odd-even relationships of the trigonometric functions.

Question No.	Answer	Detailed Explanations
6	C	The correct answer is C because by the birthday cake describes the verb placed.
7	D	The correct answer is D because, while I finish making the dough is a dependent clause that modifies the verb. It is a clause as opposed to a phrase because it contains both a subject and a predicate.
8	D	The correct answer is D because extremely expensive is an adjective phrase; extremely describes the degree to which the present was expensive.
9	A	The correct answer is A because it could stand on its own as a complete sentence with its own subject and predicate.
10	D	The correct answer is D because a relative clause is not a complete sentence and it gives extra information about the preceding noun.

Week 9

Question No.	Answer	Detailed Explanations
1	C	The amplitude of the function y=sin x or y=cos x is 1 because the maximum value of this periodic function is 1, the minimum value is -1, and the midline equation is y = 0. The amplitude is the vertical distance between the midline and the maximum or the minimum, which is 1. The trigonometric function in this problem is $y=9\ \sin(x+\frac{\pi}{2})$ which is a vertical stretch of the parent function. Therefore, the maximum value of this periodic function is 9, the minimum value is -9, and the midline equation is y = 0. The amplitude is 9.
2	B	The question asks you to find the equation of the midline of the trigonometric function $y=\frac{4}{5}\cos(2x-\pi)-5$. The amplitude of the function y=sin x or y=cos x is 1 because the maximum value of this periodic function is 1, the minimum value is -1, and the midline equation is y = 0. Therefore, the midline is the horizontal line that is exactly half way between the maximum and the minimum of the function. In the parent function, the midline is on the x-axis, and is affected by a vertical shift of the function's graph. The equation in this question has a vertical shift of 5 units downward. Therefore, the equation of the midline is y=-5.
3	B	The question asks you to find the trigonometric function that has an equation of the midline of $y=\frac{7}{2}$. The amplitude of the function y=cos x or y=sin x is 1 because the maximum value of this periodic function is 1, the minimum value is -1, and the midline equation is y = 0. Therefore, the midline is the horizontal line that is exactly half way between the maximum and the minimum of the function. In the parent function, the midline is on the x-axis, and is affected by a vertical shift of the function's graph. The trigonometric function $y=-\frac{2}{7}\cos(x-\pi)+\frac{7}{2}$ has a vertical shift of $\frac{7}{2}$ units upward. Therefore, the equation of the midline is $y=\frac{7}{2}$.
4	C	The sine function y=sin x, the parent function, has a period of 2π. The function y=a sin(bx+c) has a period of $\frac{2\pi}{b}$. In the function $y=2\ \sin(\frac{3x}{4})-5$, $b=\frac{3}{4}$. Therefore, the period of this function is $\frac{2\pi}{(\frac{3}{4})}$ which is simplified by $\frac{2\pi}{1}\times\frac{4}{3}=\frac{8\pi}{3}$.

Question No.	Answer	Detailed Explanations
5	A	The amplitude of the function y=sin x or y=cos x is 1 because the maximum value of this periodic function is 1, the minimum value is -1, and the midline equation is y=0. The amplitude is the vertical distance between the midline and the maximum or the minimum, which is 1. The trigonometric function in this problem is $y=-2\sin(x-\pi)+3$ which is a vertical stretch of the parent function by a factor of 2, a reflection across the x-axis, and an upward shift of 3 units. However, the vertical shift does not affect the amplitude. Therefore, the maximum value of this periodic function is 5, the minimum value is 1, and the midline equation is y = 3. The amplitude is 2.
6	A	The correct answer is A because disseminate means to spread or disperse.
7	B	The correct answer is B because dissidents means protesters or dissenters.
8	C	The correct answer is C because facilitate means to make easier or to enable.
9	C	The correct answer is C because re- means to do something again, as in to say again, name again, or build again.
10	A	The correct answer is A because un- means not or lacking, as in not finished, not skilled, and not friendly.

Week 9

Question No.	Answer	Detailed Explanations
1	B	If the inverse of a function is to be a function, the portion of the domain used to create the inverse function must pass the horizontal line test, or the inverse function will not pass the vertical line test. The graph of the inverse of the cosine function is found by reflecting a restricted portion of the graph of the cosine function about the line y=x. Therefore, the portion of the cosine graph that is used to create the inverse cosine graph is the interval [0,π]. The graph of the inverse function, y=cos⁻¹x is shown below. This inverse function is also referred to as y=arccos x.

$$y = \cos^{-1}x$$

Question No.	Answer	Detailed Explanations
2	C	If the inverse of a function is to be a function, the portion of the domain used to create the inverse function must pass the horizontal line test, or the inverse function will not pass the vertical line test. The graph of the inverse of the sine function is found by reflecting a restricted portion of the graph of the sine function about the line y=x. Therefore, the portion of the sine graph that is used to create the inverse sine graph is the interval $[-\frac{\pi}{2},\frac{\pi}{2}]$. The graph of the inverse function, y=sin⁻¹x is shown below. This inverse function is also referred to as y=arcsin x.

$$y = \sin^{-1}x$$

Question No.	Answer	Detailed Explanations
3	C	If the inverse of a function is to be a function, the portion of the domain used to create the inverse function must pass the horizontal line test, or the inverse function will not pass the vertical line test. The graph of the inverse of the tangent function is found by reflecting a restricted portion of the graph of the tangent function about the line y=x. Therefore, the portion of the tangent graph that is used to create the inverse tangent graph is the interval $[-\frac{\pi}{2},\frac{\pi}{2}]$. The graph of the inverse function,$y=\tan^{-1}x$ is shown below. This inverse function is also referred to as $y=\arctan x$. 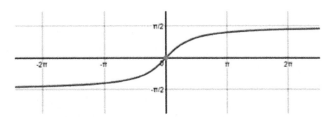
4	B	If the inverse of a function is to be a function, the portion of the domain used to create the inverse function must pass the horizontal line test, or the inverse function will not pass the vertical line test. The graph of the inverse of the cotangent function is found by reflecting a restricted portion of the graph of the cotangent function about the line y=x. Therefore, the portion of the cotangent graph that is used to create the inverse cotangent graph is the interval $[0,\pi]$. The graph of the inverse function,$y=\cot^{-1}x$ is shown below. This inverse function is also referred to as $y=\arccot x$. 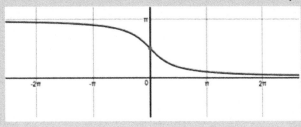

Question No.	Answer	Detailed Explanations
5	C	The graph of cosec^{-1} x is as shown below

$y = \text{cosec}^{-1}x$

For this to be a function, it should pass the vertical line test.

Thus cosec^{-1} can be defined as a function whose domain is $\mathbf{R}-(-1,1)$ and range could be any of the intervals $[\frac{-3\pi}{2}, \frac{-\pi}{2}]-\{-\pi\}$, $[\frac{-\pi}{2}, \frac{\pi}{2}]-\{0\}$, $[\frac{\pi}{2}, \frac{3\pi}{2}]-\{\pi\}$ etc. The function corresponding to the range $[\frac{-\pi}{2}, \frac{\pi}{2}]-\{0\}$ is called the *principal value* branch of cosec^{-1}. We thus have principal branch as cosec^{-1}: $\mathbf{R}-(-1,1) \rightarrow [\frac{-\pi}{2}, \frac{\pi}{2}]-\{0\}$.

Therefore, if we take the domain of cosecant x as $[\frac{-\pi}{2}, \frac{\pi}{2}]-\{0\}$, it will allow the inverse of cosecant to be a function. |
6	D	The correct answer is D because economy means the wealth and resources of a country.
7	B	The correct answer is B because legislative means relating to laws or the making of them.
8	A	The correct answer is A because analysis means an examination or investigation of something.
9	B	The correct answer is B because concepts means abstract ideas or notions.
10	C	The correct answer is C because evidence means the available facts or information.

Day 4

Question No.	Answer	Detailed Explanations
1	C	

Solve the equation $1 - \cos\theta = \dfrac{1}{2}$ if $\pi \le \theta \le 2\pi$. Begin by subtracting 1 from both sides. This gives $-\cos\theta = -12$. Next, divide both sides of the equation by -1, resulting in $\cos\theta = \dfrac{1}{2}$. The next step is the equation using the inverse of the cosine function. This gives $\theta = \cos^{-1}(\dfrac{1}{2})$. The common cosine values for Quadrants I and II are shown in the table below. In Quadrant III, the values are negative, and in Quadrant IV, the values are positive. Use symmetry to find values in those quadrants.

angle	0°	30°	45°	60°	90°	120°	135°	150°	180°
	0	π/6	π/4	π/3	π/2	2π/3	3π/4	5π/6	π
cos	$\dfrac{\sqrt{4}}{2}$	$\dfrac{\sqrt{3}}{2}$	$\dfrac{\sqrt{2}}{2}$	$\dfrac{\sqrt{1}}{2}$	$\dfrac{\sqrt{0}}{2}$	$-\dfrac{\sqrt{1}}{2}$	$-\dfrac{\sqrt{2}}{2}$	$-\dfrac{\sqrt{3}}{2}$	$-\dfrac{\sqrt{4}}{2}$

From the table and its extension, $\theta = \dfrac{\pi}{3}$ or $\dfrac{5\pi}{3}$. However, the question states that $\pi \le \theta \le 2\pi$. Thus, θ is in Quadrant III or Quadrant IV, and since the equation states that $\cos\theta = \dfrac{1}{2}$, the angle is in Quadrant IV. The angle $\dfrac{\pi}{3}$ is in Quadrant I, and the angle v is in Quadrant IV. Therefore, the answer is $\dfrac{5\pi}{3}$.

Question No.	Answer	Detailed Explanations
2	B	

Solve the equation $4\cos^2\theta = 1$ if $\pi \le \theta \le \frac{3\pi}{2}$. Begin by subtracting 1 from both sides. This gives $4\cos^2\theta - 1 = 0$. This equation is a difference of squares, so factor the left side, which gives $(2\cos\theta - 1)(2\cos\theta + 1) = 0$. Next, set each factor equal to zero: $2\cos\theta - 1 = 0$ and $2\cos\theta + 1 = 0$. Solve for $\cos\theta$: $\cos\theta = \frac{1}{2}$ and $\cos\theta = -\frac{1}{2}$. The next step is rewriting the equations using the inverse of the cosine function. This gives $\theta = \cos^{-1}(\frac{1}{2})$, and $\theta = \cos^{-1}(-\frac{1}{2})$. The common cosine values for Quadrants I and II are shown in the table below. In Quadrant III, the values are negative, and in Quadrant IV, the values are positive. Use symmetry to find values in those quadrants.

angle	0°	30°	45°	60°	90°	120°	135°	150°	180°
	0	$\pi/6$	$\pi/4$	$\pi/3$	$\pi/2$	$2\pi/3$	$3\pi/4$	$5\pi/6$	π
cos	$\frac{\sqrt{4}}{2}$	$\frac{\sqrt{3}}{2}$	$\frac{\sqrt{2}}{2}$	$\frac{\sqrt{1}}{2}$	$\frac{\sqrt{0}}{2}$	$-\frac{\sqrt{1}}{2}$	$-\frac{\sqrt{2}}{2}$	$-\frac{\sqrt{3}}{2}$	$-\frac{\sqrt{4}}{2}$

From the table and its extension, $\theta = \frac{\pi}{3}, \frac{2\pi}{3}, \frac{4\pi}{3}$, or $\frac{5\pi}{3}$. The question states that $\pi \le \theta \le \frac{3\pi}{2}$. Thus, θ is in Quadrant III. The angle $\frac{\pi}{3}$ is in Quadrant I, the angle $\frac{2\pi}{3}$ is in Quadrant II, the angle $\frac{4\pi}{3}$ is in Quadrant III, and the angle $\frac{5\pi}{3}$ is in Quadrant IV. Therefore, the answer is $\frac{4\pi}{3}$.

Question No.	Answer	Detailed Explanations
3	D	

Solve the equation $\tan(2\theta) = -1$ if $\frac{\pi}{2} \leq \theta \leq \pi$. Begin by rewriting the equation using the inverse of the tangent function. This gives $2\theta = \tan^{-1}(-1)$. The common tangent values for Quadrants I and II are shown in the table below. In Quadrant III, the values are positive, and in Quadrant IV, the values are negative. Use symmetry to find values in those quadrants.

angle	0°	30°	45°	60°	90°	120°	135°	150°	180°
	0	$\pi/6$	$\pi/4$	$\pi/3$	$\pi/2$	$2\pi/3$	$3\pi/4$	$5\pi/6$	π
tan	$\sqrt{\frac{0}{4}}$	$\sqrt{\frac{1}{3}}$	$\sqrt{\frac{2}{2}}$	$\sqrt{\frac{3}{1}}$	■	$-\sqrt{\frac{3}{1}}$	$-\sqrt{\frac{2}{2}}$	$-\sqrt{\frac{1}{3}}$	$-\sqrt{\frac{0}{4}}$

From the table and its extension, $2\theta = \frac{3\pi}{4}, \frac{7\pi}{4}, \frac{11\pi}{4}$, or $\frac{15\pi}{4}$. This means that $\theta = \frac{3\pi}{8}, \frac{7\pi}{8}, \frac{11\pi}{8}$, or $\frac{15\pi}{8}$. The question states that $\frac{\pi}{2} \leq \theta \leq \pi$. Thus, θ is in Quadrant II. The angle $\frac{3\pi}{8}$ is in Quadrant I, the angle $\frac{7\pi}{8}$ is in Quadrant II, the angle $\frac{11\pi}{8}$ is in Quadrant III, and the angle $\frac{15\pi}{8}$ is in Quadrant IV. Therefore, the answer is $\frac{7\pi}{8}$.

Question No.	Answer	Detailed Explanations
4	B	

Solve the equation $\cot\theta = 1$ if $0 \leq \theta \leq \frac{\pi}{2}$. Begin by rewriting the equation using the inverse of the cotangent function. This gives $\theta = \cot^{-1}(1)$. The common cotangent values for Quadrants I and III are shown in the table below. In Quadrant I and Quadrant III, the values are positive, and in Quadrant II and Quadrant IV, the values are negative. Use symmetry to find values in the quadrants not shown in the table.

angle	0°	30°	45°	60°	90°	120°	135°	150°	180°
	0	$\pi/6$	$\pi/4$	$\pi/3$	$\pi/2$	$2\pi/3$	$3\pi/4$	$5\pi/6$	π
cot	■	$\sqrt{\frac{3}{1}}$	$\sqrt{\frac{2}{2}}$	$\sqrt{\frac{1}{3}}$	0	$-\sqrt{\frac{1}{3}}$	$-\sqrt{\frac{2}{2}}$	$-\sqrt{\frac{3}{1}}$	■

From the table and its extension, $\theta = \frac{\pi}{4}$ or $\frac{5\pi}{4}$. However, the question states that $0 \leq \theta \leq \frac{\pi}{2}$. Thus, θ is in Quadrant I. The angle $\frac{\pi}{4}$ is in Quadrant I and the angle $\frac{5\pi}{4}$ is in Quadrant IV. Therefore, the answer is $\frac{\pi}{4}$.

Question No.	Answer	Detailed Explanations
5	C	Solve the equation $2\sin\theta + \sqrt{3} = 0$ if $\frac{3\pi}{2} \le \theta \le 2\pi$. Begin by subtracting $\sqrt{3}$ from both sides. This gives $2\sin\theta = -\sqrt{3}$. Then, divide both sides by 2: $\sin\theta = -\frac{\sqrt{3}}{2}$. The next step is rewriting the equation using the inverse of the sine function. This gives $\theta = \sin-1(-\frac{\sqrt{3}}{2})$. The common sine values for Quadrants I and II are shown in the table below. In Quadrant I and Quadrant II, the values are positive, and in Quadrant III and Quadrant IV, the values are negative. Use symmetry to find values in the quadrants not shown in the table.

angle	0°	30°	45°	60°	90°	120°	135°	150°	180°
	0	$\pi/6$	$\pi/4$	$\pi/3$	$\pi/2$	$2\pi/3$	$3\pi/4$	$5\pi/6$	π
sin	$\frac{\sqrt{0}}{2}$	$\frac{\sqrt{1}}{2}$	$\frac{\sqrt{2}}{2}$	$\frac{\sqrt{3}}{2}$	$\frac{\sqrt{4}}{2}$	$\frac{\sqrt{3}}{2}$	$\frac{\sqrt{2}}{2}$	$\frac{\sqrt{1}}{2}$	$\frac{\sqrt{0}}{2}$

From the table and its extension, $\theta = \frac{4\pi}{3}$ or $\frac{5\pi}{3}$. However, the question states that $\frac{3\pi}{2} \le \theta \le 2\pi$. Thus, θ is in Quadrant IV. The angle $\frac{4\pi}{3}$ is in Quadrant III and the angle $\frac{5\pi}{3}$ is in Quadrant IV. Therefore, the answer is $\frac{5\pi}{3}$.

Question No.	Answer	Detailed Explanations
6	D	The correct answer is D because the rest are all false. Writing to inform and explain is about choosing important facts and then developing your writing around that. Opinion has little place in an informative piece, and images can and should be used to support your paper.
7	B	The correct answer is B because facts are central to an informative piece of writing.
8	D	The correct answer is D because they all will inform the reader of something.
9	B	The correct answer is B. Informational texts should use some bullet points, so the reader can see the information more easily.
10	False	The answer is false. This type of writing would be called persuasive because the writer is using hyperbole.

Question No.	Answer	Detailed Explanations
1	B	

To find the answer, use the Pythagorean Identity, substitute, and solve for $\sin\theta$.

$$\sin^2\theta + \cos^2\theta = 1; \quad \sin^2\theta + \left(-\frac{12}{13}\right)^2 = 1; \quad \sin^2\theta + \frac{144}{169} = 1; \quad \sin^2\theta = \frac{25}{169}$$

$$; \quad \sqrt{\sin^2\theta} = \pm\sqrt{\frac{25}{169}}; \quad \sin\theta = \pm\frac{5}{13}$$

Next, use the fact that θ is in quadrant III, and the signs of the trigonometric functions change in each quadrant. The figure below shows when the signs of the trigonometric functions' signs are positive. The co-function of each has the same sign. All other trigonometric functions are negative in those quadrants.

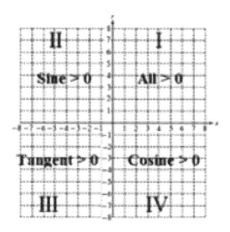

Since θ is in quadrant III, $\sin\theta < 0$, and $\tan\theta > 0$. Thus, since $\tan\theta = \frac{\sin\theta}{\cos\theta}$, so $\tan\theta = \dfrac{-\frac{12}{13}}{-\frac{5}{13}} = \frac{12}{5}$.

Question No.	Answer	Detailed Explanations
2	C	

To find the answer, use the Pythagorean Identity, substitute, and solve for $\cos\theta$.

$$\sin^2\theta + \cos^2\theta = 1 \, ; \; (-\tfrac{7}{25})^2 + \cos^2\theta = 1 \, ; \; \tfrac{49}{625} + \cos^2\theta = 1$$
$$\cos^2\theta = \tfrac{576}{625} \, ; \; \sqrt{\cos^2\theta} = \pm\sqrt{\tfrac{576}{625}} \, ; \; \cos\theta = \pm\tfrac{24}{25}$$

Next, use the fact that θ is in quadrant II, and the signs of the trigonometric functions change in each quadrant. The figure below shows when the signs of the trigonometric functions' signs are positive. The co-function of each has the same sign. All other trigonometric functions are negative in those quadrants.

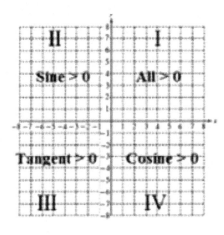

Since θ is in quadrant II, $\cos\theta < 0$. Therefore, $\cos\theta = -\tfrac{24}{25}$.

Question No.	Answer	Detailed Explanations

3   **D**

To find the answer, use the Pythagorean Identity, substitute, and solve for $\cos\theta$.

$$\sin^2\theta + \cos^2\theta = 1; \ (\tfrac{3}{5})^2 + \cos^2\theta = 1; \ \tfrac{9}{25} + \cos^2\theta = 1$$
$$\cos^2\theta = \frac{16}{25}; \ \sqrt{\cos^2\theta} = \pm\sqrt{\frac{16}{25}}; \ \cos\theta = \pm\frac{4}{5}$$

Next, use the fact that θ is in quadrant I, and the signs of the trigonometric functions change in each quadrant. The figure below shows when the signs of the trigonometric functions' signs are positive. The co-function of each has the same sign. All other trigonometric functions are negative in those quadrants.

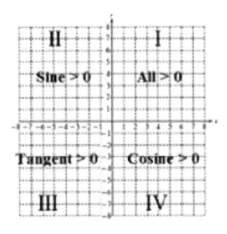

Since θ is in quadrant I, $\cos\theta > 0$. Therefore, $\cos\theta = \frac{4}{5}$.

Question No.	Answer	Detailed Explanations
4	B	Note that $\cos 75° = \cos(30° + 45°)$. Using the formula for the cosine of a sum, which is $\cos(\alpha + \beta) = \cos \alpha \cos \beta - \sin \alpha \sin \beta$, substitute the values for of 30° and 45°. Therefore, $\cos 75° = \cos(30° + 45°) = \cos 30° \cos 45° - \sin 30° \sin 45°$. The table below provides these values.

angle	0°	30°	45°	60°	90°
	0	$\pi/6$	$\pi/4$	$\pi/3$	$\pi/2$
sin	$\frac{\sqrt{0}}{2}$	$\frac{\sqrt{1}}{2}$	$\frac{\sqrt{2}}{2}$	$\frac{\sqrt{3}}{2}$	$\frac{\sqrt{4}}{2}$
cos	$\frac{\sqrt{4}}{2}$	$\frac{\sqrt{3}}{2}$	$\frac{\sqrt{2}}{2}$	$\frac{\sqrt{1}}{2}$	$\frac{\sqrt{0}}{2}$

From the table, $\cos 30° \cos 45° - \sin 30° \sin 45°$

$$= \frac{\sqrt{3}}{2} \times \frac{\sqrt{2}}{2} - \frac{1}{2} \times \frac{\sqrt{2}}{2}$$

$$= \frac{\sqrt{6}}{4} - \frac{\sqrt{2}}{4}.$$

Question No.	Answer	Detailed Explanations
5	A	Note that $\tan 15° = \tan(45° - 30°)$. Using the formula for the tangent of a difference, which is $\tan(\alpha - \beta) = \frac{\tan \alpha - \tan \beta}{1 + \tan \alpha \tan \beta}$, substitute the values 45° and 30°. Therefore, $\tan 15° = \frac{\tan 45° - \tan 30°}{1 + \tan 45° \tan 30°}$. The table below provides these values.

angle	0°	30°	45°	60°	90°
	0	$\pi/6$	$\pi/4$	$\pi/3$	$\pi/2$
tan	$\sqrt{\frac{0}{4}}$	$\sqrt{\frac{1}{3}}$	$\sqrt{\frac{2}{2}}$	$\sqrt{\frac{3}{1}}$	∎

From the table, $\dfrac{\tan 45° - \tan 30°}{1 + \tan 45° \tan 30°} = \dfrac{1 - \frac{\sqrt{3}}{3}}{1 + 1 \times \frac{\sqrt{3}}{3}} = \dfrac{3 - \sqrt{3}}{3 + \sqrt{3}}.$

Question No.	Answer	Detailed Explanations
6	C	The correct answer is C because plagiarism is defined as stealing from a source without giving that source credit.
7	A	The correct answer is A. When you cite your sources correctly you inform the reader of where you gathered the information from.
8		Paraphrasing is borrowing the idea, but not the words and yes, you do need to cite it.
9	D	The correct answer is D because these are the defining characteristics of a good research statement.
10	D	The correct answer is D because those are all different styles.

Notes

BLOGS

7 Signs You are on the Right Track to Ace High School

High school is not a cakewalk for everyone. In order to do well, you need to know what your teacher's expect, how to reach those expectations, and how to get the best grades you can prior to applying for college.

How Top Performers Reach Academic Goals

It is not meant to be a secret; reaching your goals takes planning, hard work, and self-motivation. People who can do this have worked hard to make a plan and stick to it. Setting goals and making a list of steps needed to reach that final goal will help keep you motivated to get there.

Write Your Own Notes

Reading a textbook is fine but writing the information down will help you to remember it in a way that makes sense to you. When you take your own notes, whether it be from a textbook or a lecture, you will remember more of the information.

Note Taking Tips

- Write down the key points.
- Do not overdo it. You will not be able to write things word for word.
- Ask questions. If you feel you missed something that may be important, ask for it to be repeated. Or if you need to clarify a point, ask.
- Organize your notes when you have time. When you do a review in the evening, go through your notes, and make sure you can still understand them. You may want to expand on them and organize them into charts or diagrams.

Seek Help

Successful people do not do everything on their own. If you want to be successful you need to make sure you understand all the course material.

If you are struggling with information or a concept, make sure to get help so that you can understand it before a test or exam.

You can ask other students in the class, ask your teacher for extra help, or get a tutor.

Peer Discussions

When you are learning new material, it is helpful to discuss it with others. Like writing things down, being able to discuss it will help you to retain the information.

If you can get your point across or teach the concept to your peers, it shows that you have a clear understanding of the information.

Time Management

To be successful, you need to know how to manage your time effectively. Keep your goals written clearly where you will see them every day. Make a calendar of things you need to do to attain those goals. Follow your schedule as closely as you can to make sure you are prepared in advance of upcoming exams. That will give you time to go back and review everything and ask for any clarification where needed.

Sufficient Practice and Mock Exams

Make sure you practice the skills you are learning and frequently review the information you must remember. The more you practice and review, the more easily you will recall the information.

You can often find practice tests online or you can ask your teacher for an old test to use for review and practice.

In a Nutshell

To ace high school, you need to be prepared, have a goal, and work hard to get there. Do not be afraid to ask someone to help you to set up a schedule and to help you stick to it!

How To Master The Art Of Essay Writing For High School Students

By now you have probably written a lot of essays. Whether you love them or hate them, there is no denying that school assessments are leaning more and more toward this long-form style of writing, so mastering it is crucial. Whether you are a current or soon-to-be high-schooler, there is an art to essay writing that, if properly honed, can help you ace it every time.

Here are our top tips for essay writing:

Understand essay and paragraph structure

One of the most simple yet important steps of essay writing is sticking to the correct structure. The body of the essay should have an introduction that states your argument, well-researched body paragraphs to support, and a conclusion that sums up the essay. Each paragraph should tackle an individual idea, and be presented in a logical sequence to support your overall argument.

With the correct structure, an essay's argument will be easy to follow and always refer back to the question.

Create a plan

While the structure is a format that can be used for every essay, the plan uniquely organizes your research for each specific essay. Start with brainstorming and mind mapping, and once you have a general flow of ideas move onto more in-depth research on each point.

Finding the right sources is a skill to be practiced, but once you are researching in the correct place it should be easy to create an essay plan for any subject. Organize information and key points to make it simple to refer back to later.

Edit and review

Polishing your essay is an art, and the crucial final step in writing. This is the time to make sure that the overarching questions are answered concisely, sentences are well-structured, and there are no pesky typos hiding in there. In an exam setting, it can be harder to edit due to time constraints, but if you are writing from home it is even a good idea to leave it overnight and revise with fresh eyes in the morning.

Whether you are new to the essay or confident in your abilities, honing the basics is always a good idea. With a bit of practice, you will be acing essays and getting top grades. Good luck!

How Reading Fluency Will Help You Ace the SAT and ACT

Reading fluency will help you improve your SAT and ACT scores. The first important step is to find books that you enjoy. Reading can be boring if you do not have something that engages you. Once you find the kind of books you enjoy, read as much as you can.

SAT and ACT

The more you read, the more your reading fluency will improve. The verbal section of the SAT is based on reading and understanding passages. Since it is a timed test, you will need to be able to read at a certain pace to get all of the questions answered.

Even if you have a high level of understanding, that does not help much if you read at a slow pace. The more you read every day, the faster you will be able to read and comprehend the passages on the test. When you are a more efficient reader, you will have more time to answer the questions on the test.

The ACT measures college readiness and to be accepted into a college, you need an acceptable level of reading fluency as well.

Strategies Used to Evaluate Reading Fluency

To assess fluency, evaluators often used timed oral reading tests and then ask comprehension questions at the end of the reading.

On a standardized evaluation, there will be reading comprehension and vocabulary sections that are used to assess the applicant's fluency.

Improve Your Reading Fluency

The best thing you can do to improve your fluency is to read every day. The more you read, the more fluent you become.

Read to yourself and then put the book down and see if you can summarize what you have read.

Tips for the Test

- Identify the mistakes you make most often when you are reading so you can focus on improving in that area.
- Practice skimming passages.
- Skip difficult questions and return to them later.
- Use the process of elimination when needed.
- Practice!

The SAT

On the SAT, the first section is reading comprehension and vocabulary. You will be provided five passages that you will need to read and then you will have 52 multiple choice questions to answer based on the reading.

This section of the test will evaluate your command of evidence and words in context.

To get ready, you should practice reading passages similar to what will be found on the SAT.

The ACT

The American College Test is like the SAT and will test the same skills. To do well on this test, you will need to be able to read fluently. Reading fluently will help to improve both comprehension and vocabulary.

Extra Tips

Though it can be more difficult to improve fluency at an older age, it is an attainable goal.

Read as much as you can and read out loud when you can. It is also a good idea to listen to other readers and pay attention to their pace. Audiobooks are a good way to do this.
Practice will improve your skills.

Writing a Killer Statement of Purpose

When you apply for college, you will have to write a statement of purpose. This is an essay that will showcase your writing abilities as well as who you are. The admissions committee will use this essay to determine the best candidates to attend their college.

The statement of purpose describes who you are, why you are applying to the college, why you are a good candidate, and what you hope to achieve in your future. The committee will use this piece of writing to assess your writing abilities.

Tips for your Statement of Purpose
- Make sure it is free of spelling and grammatical errors.
- Use clear and concise language.
- Do not use cliches or informal language.
- Make sure the tone is positive and confident.

How to Make it Persuasive
In this piece of writing, you are showcasing your talents and selling yourself. You need to persuade your audience that you are the best candidate.

For a statement of purpose, you have one page to sell yourself. Your language must be concise. If your statement of purpose is much longer than a page, it may seem like your writing lacks focus and clarity.

The Process
Start writing early so that you will be done well ahead of the deadline. This gives you time to look over your writing and make any improvements needed.

Make sure that you understand the question that is being asked so that you write in an appropriate manner to respond to the prompt.

Do some self-reflection before you start writing. Make notes about what makes you unique and why you think you are a good candidate.

Do some research on the school to see what it is known for and assess how you fit into that kind of environment. You want to persuade the admissions board that the school is a good fit for you but that you are also a good fit for the school.

Once you have written a draft, go through and edit it. It is also a good idea to have someone else read it over for you to get some feedback.

Before you Submit it
Before you are finished with your statement of purpose, be sure to read it over to ensure there are no errors. Proofreading is an important step in the writing process and even more important now that you are competing against others for a spot in the college.

Try to read it from the admissions perspective and see if you would have any unanswered questions!

Why You Should Master The Art Of Persuasion In Public Speaking As A Student

Whether you plan to try out for debate this year, or just want to converse more convincingly, working on your persuasive speaking is always a good idea. Understanding the way we compose and deliver persuasive language may help you win a discussion with your parents, but there are even more benefits than you may expect. Here are our top reasons why you should work on mastering the art of rhetorics:

Improve reading fluency

You may not expect it, but by improving your grasp of rhetorics (the art of persuasive speaking), you will also improve your reading skills. Being able to convincingly read passages out loud with the correct emphasis improves reading ability overall.

To see where your reading fluency is at, try reading a passage out loud to the [Lumos Reading Fluency Analyzer].

Deliver crisp speeches and presentations

Do you dread classes that assess you on public speaking? It may be scary, but giving speeches is a skill that is easily improved. Practice reading passages aloud, and soon you will start to naturally create emphasis and pace that supports whatever you are speaking about. This means that when an assessment comes up, you will be well-practiced, not shy, and ready to ace it!

Prepare for moot court or MUN debates

Once you have basic public-speaking mastered, you can start to plan where your new skill may take you. Moot court, MUNs, and your school debate team are all great ways to flex your skills of persuasion and are of course great activities to have on your college application.

Sometimes it is easy to forget that reading, writing, and speaking are all connected, which is why mastering rhetorics is an important skill to connect them all. By improving your composition and delivery of language, you are honing advanced language skills that will last you a lifetime.

How To Ace In Competitive Exams (ex: SAT, PSAT/NMSQT, ACT)

It doesn't matter which test you are studying for (the SAT, the PSAT/NMSQT or the ACT), acing it is an important step in the road to your dream college. Preparing for an exam can feel daunting, but by breaking the process down it becomes a more manageable task. Here are some tips to perfect your study process and ace that upcoming exam!

- **Take AP classes**

Because many of the exams are directly related to concepts and critical thinking skills learned in the classroom, these challenging classes will inevitably help you prepare for exams. It is shown that students that did well in two AP classes had generally much higher SAT scores.

- **Decide on a target score**

You are probably taking one of these tests for a specific reason: to get into your dream college, course, or scholarship. Which means you most likely already know, at least roughly, what score you need in your tests. By having a clear goal, you can then plan to create a plan to balance your weaknesses and strengths and increase your overall score.

- **Set up a personalised practice schedule**

Hopefully, you have left a lot of time to study, as it's much more manageable to break the exam prep into weekly goals. You probably have a strong understanding of your own personal strengths and weaknesses, but doing a few practice questions will help confirm this. Once you have identified the main areas to work on, you can allocate the needed time. The College Board and Khan Academy have excellent resources and quizzes to get you on the right track.

- **Take a full-length practice exam**

Once you've been studying for a while, the best thing you can do is take a full-length practice test (or more!) It will get you used to the long exam and test structure, and help you gain confidence working under pressure. You can either print it out and practice with paper and pencil (like the actual test will be), or do it online for instant results and more interactive feedback. Or, why not do both?

The main thing to know is that with the right plan and diligence, you have nothing to worry about... you're going to ace it!

It's Not Too Early To Be Thinking About College! How To Research Your Dream School.

While college applications may still be a few years away, you can give yourself a head start by starting to think about it now. With a clear intention of where you see yourself at college, you can use these high school years wisely to increase your chances of getting in. These are some of the general steps you can take to find your dream school.

1.   Decide on a career industry

While you may not know exactly what job you want in the future, you probably have an idea of what industry you want to go into. Law? Architecture? The arts? Every college will have different areas of specialty, so realizing your career goals will make researching schools easier.

Of course, it is okay if you are not quite sure of what path your future career will take... this is totally normal! In this case, it is worth looking for a college that has a wide range of class options. This will let you keep your options open and make a strong career choice in the future once you've had some more time and experience.

2.   Realize your priorities

A lot of schools may sound perfect on paper, but college is such an important life experience that it's important your choice matches you as a person. Questions to ask yourself include: how far do you want to be away from home? Are there any clubs or societies that match your interests? How many students do they have? Even researching the weather in that part of the country may have a strong impact on your decision.

3.   Do your research

Once you know what you value in a college, it's time to research. Read the schools' websites, find blogs from students, visit potential campuses if you can. You will probably find that the more you learn about a college, the more you either like or dislike it. You still have a lot of time, but the more comfortable you are with your options now, the easier the application process is going to be in a few years.

4.   Make a shortlist

Notice it says a shortlist, rather than choosing one specific college. Sure, your favorite choice can sit at the top of the list, but right from the beginning, it is important to be open-minded about college admissions. Over the next few years, your opinions, grades, and interests will all affect the college you eventually choose. Having a list now, however, means you can research unique admission factors and make sure you're ticking them off. If your dream school values certain athletic or charitable pursuits, for example, you have a lot of time to make sure you're ready to apply.

College is such an exciting time in life and definitely something to look forward to. It's never too early to start researching your dream school... good luck!

Additional Information

What if I buy more than one Lumos Study Program?

Step 1

Visit the URL and login to your account.
http://www.lumoslearning.com

Step 2

Click on 'My tedBooks' under the "Account" tab.
Place the Book Access Code and submit.

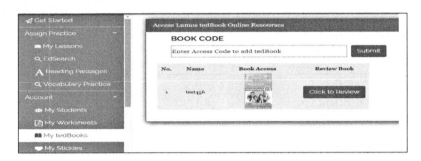

Step 3

To add the new book for a registered student, choose the
○ Existing Student button and select the student and submit.

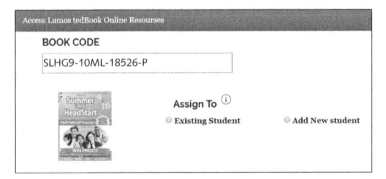

To add the new book for a new student, choose the ○ Add New student
button and complete the student registration.

Lumos tedBooks for State Assessments Practice

Lumos tedBook for standardized test practice provides necessary grade-specific state assessment practice and skills mastery. Each tedBook includes hundreds of standards-aligned practice questions and online summative assessments that mirror actual state tests.

The workbook provides students access to thousands of valuable learning resources such as worksheets, videos, apps, books, and much more.

Lumos Learning tedBooks for State Assessment	
SBAC Math & ELA Practice Book	CA, CT, DE, HI, ID, ME, MI, MN, NV, ND, OR, WA, WI
NJSLA Math & ELA Practice Book	NJ
ACT Aspire Math & ELA Practice Book	AL, AR
IAR Math & ELA Practice Book	IL
FSA Math & ELA Practice Book	FL
PARCC Math & ELA Practice Book	DC, NM
GMAS Math & ELA Practice Book	GA
NYST Math & ELA Practice Book	NY
ILEARN Math & ELA Practice Book	IN
LEAP Math & ELA Practice Book	LA
MAP Math & ELA Practice Book	MO
MAAP Math & ELA Practice Book	MS
AZM2 Math & ELA Practice Book	AZ
MCAP Math & ELA Practice Book	MD
OST Math & ELA Practice Book	OH
MCAS Math & ELA Practice Book	MA
CMAS Math & ELA Practice Book	CO
TN Ready Math & ELA Practice Book	TN
STAAR Math & ELA Practice Book	TX
NMMSSA Math & ELA Practice Book	NM

Available

- At Leading book stores
- www.lumoslearning.com/a/lumostedbooks